Gary Fradin

How to Be a Patient

A Manual to Help Patients Get Better Care with Less Risks and at Lower Costs

Gary Fradin

Gary Fradin

ISBN: 978-0-359-30191-1

**Nothing in this book should be interpreted as medical advice.
The information presented here is for illustration purposes only.**

Gary Fradin

Table of Contents

Preface

Never write an article that you can summarize in a tweet;
Never write a book that you can summarize in an article;
Only write a book that needs to be read more than once.

I read this aphorism about modern communication somewhere and thought about it constantly as I wrote this book.

You can't become a wise patient by reading a few tweets or googling a few articles.

Instead, it's an educational process. Learn the fundamentals first, then practice making wise decisions by addressing your own, hopefully small, medical needs over time.

Becoming a wise patient is, I submit, a learnable skill. Those who acquire it will predictably enjoy better medical outcomes with less risk and at lower costs..

You need a guide, though, to help you navigate the decision making process.

I offer this book for that purpose. It will tell you how to prepare for your doctor's appointments, what to discuss in sick visits, how to understand well visits and how to do your own research.

It won't tell you what care to get!

Instead, it focuses on how to decide what care to get.

I hope you will read it first as an overview, then refer back to specific sections as your medical needs require. Feedback from initial readers indicated as much.

Feel free to share this information with your doctors. Medicine is a team sport with the best teams consisting of well informed patients and wise, caring doctors.

More people than I can count helped me develop the ideas here, most unknowingly. They were the students who listened to my lectures, read my previous books, asked tough questions and provided feedback. I'd like to thank them all.

And I'd especially like to thank the following for providing insights, comments and most importantly, support and encouragement as I developed these ideas. In no particular order: Tom Hamel, Bill Stuart, Michele Boyer, Kerry Stefano, Isaac Fradin, Sue Donahue, Jan Ruderman, David Rubin, Jeff Rich, Andy Roberts, Larry Croes, David Mudd, Francis Russo, Jeff Krawczyk, Laurie Moran, Don Poulin, Judy Besse, Lois Drukman, Chris Hammond, Bob Landry, Todd MacDonald and especially Joe Fraiman for his excellent and detailed written feedback. I'm sure I left some people out and apologize for that.

Last and foremost (!), I offer a special thanks to my wife, Marjorie, for putting up with me through years of research, confusion, writing and life in general. I often hear how people's lives would have been less rich, productive and satisfying without the love and support of their spouse, and I know that's the case for me. I certainly hope to spend many more satisfying, fulfilling and confusing years with Marjorie.

In the end, though, since I wrote this book, all errors, poor research and crummy explanations are entirely my own.

Introduction
The end of amateurism

Great cooks take cooking classes, good cooks read cook books and mediocre cooks rely on personal experience.

Ditto for artists, writers, musicians, craftsmen and more.

Today we get advanced training in lots of previously routine, ordinary activities. We've come to accept that the more training we get in a field, the better we get at it.

Except for the most important and dysfunctional thing we do, receive medical care.

Most important? Your life may depend on your decisions.

Most dysfunctional? We annually waste hundreds of billions of dollars on ineffective or harmful care.

No Patient 101 courses exist in high schools or colleges, no introductory educational programs in doctor's offices, no patient training in hospitals, no overview modules from HR in large self-funded companies, not even any specific training for doctors or nurses in medical schools.

What would a patient education course teach? What body of knowledge, in other words, would a great patient acquire? Here's a short list of topics I'd recommend:

- **How to choose care that works** i.e. benefits patients based on high quality scientific studies. Only some of medical care is based on science these days according to many estimates; the rest is guesswork and hunches.[1]

  That's why great patients routinely get opinions from different physicians with different treatment orientations.

  We'll discuss why and how to do this in Chapter 2.

- **How to avoid care that doesn't work** or doesn't work well based similarly on scientific studies, things like extended release niacin to reduce heart attack risks, joint lavage to reduce knee pain, beta blockers to prevent heart attacks and so one. I'll provide case studies of these and others in this book.

Patients too often trust theories, not hard evidence, thinking 'it should work because the underlying biology says so.'

Indeed, I've heard countless people in my classes say, when I present a case study showing a particular treatment doesn't benefit patients, 'the study must be wrong.'

I'll explain what good, reliable scientific evidence is in Chapter 1 and elsewhere, and show why relying on it can improve your outcomes and reduce your risk of harm.

Not to mention reduce your treatment costs.

- **How to identify care that's overused** even if studies show that it works on a well-defined patient group, because we know that overuse accounts for about third or so of medical care in the US today,[2] things like

    Imaging for eye disease, overused according to one large study about 74% of the time,[3]

    Antibiotics for upper respiratory and ear infections, overused about 98% of the time,

    Cardiac stress tests, overused about 19% of the time and representing over $2 billion in annual waste, among others. I'll discuss this issue more in Chapter 1.

- **How to choose among care alternatives** because you have treatment options almost all the time.[4] Surgery or physical therapy? More aggressive surgery or less? Physical therapy or medications? Watch-and-wait or treat now? How does a wise patient decide? We'll discuss that in Chapter 2.

Studies consistently show that better informed patients – that means better informed about the likely outcomes and treatment options – tend to choose less invasive, less risky and typically therefore less costly interventions more frequently than do less well-informed patients, the amateurs.

- **How to use screening tests effectively** because some tests generate unreliable information while others are tremendously beneficial. How does a wise patient decide which screening tests to have and how frequently?

  Related to this, at what test result should *you* do something, at blood pressure of 130/80 ... or 145/85 ... or 160/90 for example?

  I'll discuss in Chapter 3 how different organizations make different, sometimes contradictory recommendations but virtually all leave out critical mind-body or emotional factors. Should a happy, socially active, optimistic, financially well-off, athletic fellow take medications at the same blood pressure as a depressed, impoverished, lonely one? Read Chapter 3, consider the implications, then discuss your own situation with your doctor.

- **How to choose the best specialist and hospital** for fairly obvious reasons. I'll recommend a good, but not perfect, way to identify the best: *first* decide which treatment alternative you prefer, which specific type of surgery for example, and *then* determine which surgeons perform it the most. It's the best rule-of-thumb available for determining specialist quality, though it's not perfect.

  See Chapter 2 again for a more detailed explanation then consider how applying this to your own physician and hospital decisions might benefit you.

- **How to understand a medical study, article or ad** because they're omnipresent in today's media and everyone relies on Dr. Google to some extent.

  As a quick introductory comment, it's way more difficult to read a medical article critically than most people think.

And the headlines, the part that too many people rely on, may misrepresent critical nuances that can affect your own care outcomes.

I'll offer some article-reading advice in Chapter 4.

My hypothetical Patient 101 course wouldn't focus on medical prices or insurance coverage. I've never heard anyone say 'I won't give my child necessary care until next Open Enrollment when we can switch to a plan that covers it.'

But I have heard people say 'I can't afford all this medical care so I'll only get some', which means they have to choose which to get and that takes us back to the care quality and wise patient issues.

Why well informed patients cost less

Wise patients – the professionals in my terms – know 3 things that amateur patients generally don't.

- How to identify and avoid unnecessary, ineffective and overused medical care, perhaps 30% of all spending. [5]

- How to identify, explore and compare treatment options, available to patients about 85% of the time. [6]

  How to identify and choose better quality care providers. Better quality means fewer errors and hospital readmissions, shorter hospital stays and a quicker return to think.

- And three leads to lower medical expenditures and all three together lead to much lower medical spending. That's why learning to get better care with less risk will save patients and payers money.

About this book

This book is a manual, not an academic piece or medical advice. My interest here is not *what* medical care you choose but *how* you choose it.

I hope people will read it initially as an overview and then refer back to various sections as needed for their own medical care.

Reviewers of early drafts said as much with comments like

- 'I'll ask the questions in Chapter 2 at my next doctor's visit.'
- 'I didn't know about the information sources in Chapter 1. I'll check them out.'
- 'I want to show Chapter 4 to my wife. She does tons of online research and could use your pointers.'

And the most frequent comment from both lay folks and medical professionals: I didn't know that!

A message for CEOs and CFOs

Corporate attempts to control employee healthcare expenses over the past 40 years have generally failed. We know this because corporate healthcare premiums (employer + employee contributions + deductibles) have inflated much faster than the overall Consumer Price Index since the 1970s.

The fundamental reason is not necessarily inappropriate plan designs.

Instead, it's the attempt to solve clinical problems with financial or insurance tools, things like

- **Higher deductibles** to reduce wasteful spending but without defining waste or suggesting ways for your employees to differentiate high from low quality care.
- **Medical price lists**, to which amateur patients may respond 'I want the higher priced care because it's probably better'.
- **Wellness programs** that reward your current healthy employees financially but make the least well employees, i.e. your most expensive medically, feel badly so they don't participate.
- **Health risk assessments**, financially incentivized, meant somehow replace a primary care physician's advice.

- **Tax saving programs** – HSAs, HRAs, FSAs for example - that confuse participants and don't improve patient outcomes, and more.

We impose all this on employees who often lack critical medical decision making skills: some 88% of Americans are medically illiterate according to the US Department of Health and Human Services. [7] Illiterate means 'hasn't been trained' not 'stupid'.

'Medically illiterate' also means unable to estimate the likely benefits and risks of medical care.

Imposing financial incentives on this group can't possibly generate satisfying results either for you or them and it hasn't.

But there's an alternative approach: expanding employee medical literacy through a serious and well organized education program. Consider the potential impact on your utilization rates from this conclusion to the 2012 Patient Preferences Matter report, jointly authored by Dartmouth medical and business school professors: [8]

> Well informed patients consume less medicine – and not just a little bit less, but much less.

And this observation from Dr. Sandeep Jauhar in his autobiographic book Doctored, largely a description of his years overtreating patients:

> Better informed patients might be the most potent restraint on overutilization.

To make your health insurance program work - reduce corporate medical spending, decrease unnecessary utilization and help your employees get or remain healthy – you need to include employee education about care quality. I'll outline that content in this book.

Why I don't discuss health insurance in this book

Health insurance and medical care quality are two separate and largely unrelated bodies of knowledge. Attempting to integrate them in one text leads, in my opinion, to confusion rather than clarity.

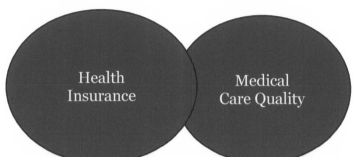

Understanding health insurance involves learning about policy provisions, deductibles, referral requirements, in-and-out of network constraints, pricing, coverage mandates and a host of related issues. These govern how and under what conditions an insurance carrier pays for medical care.

Understanding care quality involves learning about medical outcomes, care overuse, treatment options, disease risks, medical studies and similar related issues. These describe how well medical care actually works, how likely a patient is to benefit from it and what treatment options exist.

Someone can be well informed about health insurance and poorly informed about care quality and vice versa.

Interestingly, the 'well informed about care quality' patients tend to use less medical care and cost less; they focus on their likelihood of benefiting from a specific intervention.

Meanwhile, the 'well informed about health insurance' patients may use more medical care and cost the most; they understand how to maximize insurance payments.

Given this potential confusion and conflict, I'll try to do one thing well, not two confusingly. I'll focus on medical care quality in this book.

About me

I'm not a doctor and I don't give medical advice.

Instead, I'm an economist. I study *how well* medical care works and why.

I teach health insurance continuing education classes for a living, focused largely on medical care quality and related issues. I've given over 500 lectures and written 6 books on the topic since 2002.

All that study has convinced me that **care quality responsibility rests primarily on each individual patient's shoulders.** The best way to ensure that you get high quality medical care is to engage in the *how likely am I to benefit and be harmed, and how* discussions introduced here with your caretakers.

I'm equally convinced that patients can't turf that process to others – doctors, hospitals or insurance carriers for example. I'll explain why in the next 100 or so pages.

I received my formal training in City and Regional Planning at Harvard. There we learned how to evaluate investments – understand the context, measure the expected benefits, allow for downside risk and compare options.

I simply apply the same approach to medical care. I don't understand why anyone would do otherwise.

This book took me 10 years and two months to write: 10 years of research, lectures and student feedback to develop the ideas, then two months to write it all down.

I hope it helps when next you need medical care.

Chapter 1: How to Prepare for Your Doctor's Appointment

This chapter focuses on patients who see physicians for scheduled appointments either for existing, known medical conditions or for annual physicals.

Let's start with this observation from Dr. Atul Gawande, named in 2018 to run the Berkshire Hathaway-Amazon-JP Morgan healthcare initiative:

> the ideal modern doctor should be neither paternalistic nor informative but rather interpretive, helping patients determine their priorities and achieve them [9]

Unpack Gawande's comment:

- 'Paternalistic' means making decisions for the patient.
  Ideal modern doctors don't want to decide for patients, in part because so many medical decisions fall into the gray area between *definitely doing something* and *definitely not.* Doctors also realize that patients today generally have multiple treatment alternatives: surgery vs. physical therapy for example or medications vs. surgery. Similar patients with similar medical problems can make different treatment decisions and all be right.

Doctors increasingly think patients should make these decisions, not physicians. Very different from 50 years ago.

- 'Informative' means act like an encyclopedia. Doctors don't want to perform this function in the internet age; it's a waste of their time. be right.

  Today's average doctor visit lasts about 15 minutes. That's insufficient for doctors to listen to patients, diagnose the ailment, outline treatment options, understand the patient's treatment preferences *and* provide a literature review and summary. Gawande wants doctors out of the routine information dissemination business.
  This chapter discusses the informative function in detail.

- 'Interpretive' means apply scientific facts and the physician's own treatment experiences to a specific patient's needs.

  Ideal modern doctors understand that wise medical decisions combine facts and art. The facts come from studies and journals, the art comes from your physician's experience.

  That's why physicians 'practice' medicine and why you don't get all your care from a computer and robot.

For modern doctors to act ideally, in Gawande's terms, they need patients to learn the most important facts prior to appointments so they arrive ready to interpret.

This chapter explains how to do that. It's actually a pretty easy, pretty quick and immensely eye-opening process.

Three information sources

I strongly encourage patients to use three reference sources – Cochrane, the US Preventive Services Task Force and ChoosingWisely – and only these unless you've been trained in how to read and understand a medical article. That's Chapter 4 of this book and it's more complicated and difficult than most people think. Even then, relying only on these three sources will serve you extremely well.

I realize that this is controversial. Based on my own research, analysis and experience, however, I contend most patients, most of the time, can get all the information they need from these three sources.

Why these three? They all

- Focus on patient outcomes, addressing the 'how well does it work?' question.
- Receive no industry funding so face no financial conflicts of interest or report biases.
- Provide objective and transparently developed information.
- Are practical and easy-to-use, and
- Have a Search Box so you can look up your own specific situation. (Caveat: the Search Boxes work *pretty* well, not *wonderfully* well so you may need to try a few different word combinations. That process generally adds a few seconds to your task.)

I worry about two main problems using other sources, some of which may be very good.

Some focus on 'how' medical care works, the nuts-and-bolts of liver function for example, rather than 'how well' it works, the outcomes patients can expect from various interventions.

Care that *should* work, based on 'how' analyses, sometimes doesn't according to outcome studies. In that case, based on my own anecdotal evidence, people tend to believe the theory, not the studies. This is an error in my opinion, sometimes a very big one.

I always encourage people to focus on the outcomes of medical interventions, not the underlying biology. Trust the studies, assuming, of course, that they're well done. These three sources ensure that.

Second, some commercially supported sites may face potential conflicts of interest between their advertising and reporting functions. WebMD, for example, received about $428 million from biopharma and medical device advertising and sponsorships in 2016, about 60% of their operating costs. [10]

The resulting potential for information bias makes me uncomfortable.

The three sites I use do not face these potential problems.

Cochrane

I use Cochrane as my first medical information source.

Cochrane, named after Oxford University researcher Archie Cochrane, provides a comprehensive library of high-quality evidence so doctors can give good advice and patients can make informed decisions.

It accepts no commercial or conflicted money but receives income from subscriptions to its library – the summary write-ups are free and available without a subscription - and grants from, among others, the US National Institutes of Health, the UK National Institute of Health, the Australian National Health and Research Council, the Canadian Institutes for Health Research, the Robert Wood Johnson Foundation and the Bill and Melinda Gates Foundation.[11]

As a brief editorial comment, when Bill Gates chooses to fund something, pay attention.

Cochrane, formerly known as the Cochrane Collaborative, is highly regarded in the medical research community. Here are a few comments about it:

- From Duke University's Medical Center Library: "The Cochrane Library contains high-quality, independent evidence to inform healthcare decision-making. It includes reliable evidence from Cochrane systematic reviews and a registry of published clinical trials. The methodology used to create the Cochrane reviews is recognized as the gold standard for developing systematic reviews."[12]
- From Nature's Editor-in-Chief, L.A. Harvey: "The Cochrane Collaboration has developed stringent methodologies to reduce bias and ensure trustworthy interpretation of the evidence." [13]

- From the Lancet: "The Cochrane Collaboration...rivals the Human Genome Project in its potential implications for modern medicine."[14]

Alan Cassels' 2015 book title, **The Cochrane Collaboration: Medicine's Best Kept Secret**, summarizes both the public awareness of Cochrane (low) and impact (high).

Cochrane and many (most?) medical researchers believe that the best medical evidence comes from controlled comparative studies. At the simplest level, these compare two identical groups of people, one of which gets the medical care and the other of which does not.

Here's a simple graphic description of a controlled comparative study of Drug X designed to prevent heart attacks. People in the Treatment Group took it; people in the Control Group did not.

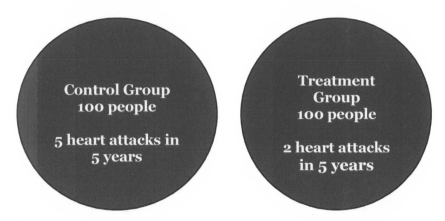

Researchers conclude from this trial that Drug X prevented 3 heart attacks per 100 people in 5 years. Pretty straightforward.

But sometimes the researchers goof. Maybe they biased their results by choosing a non-representative population. Maybe they organized the two groups inappropriately. Maybe they messed up the data.

For these reasons among others, some experts like Dr. Marcia Angell of Harvard Medical School suggest that wise patients always find *more than one* study showing roughly the same results before deciding to have a medical intervention. [15]

I heed Dr. Angell's advice because of her unique position in American healthcare: she was Executive Editor of the New England Journal of Medicine, the most prestigious peer-reviewed medical journal in the world, for about a decade, then Editor-in-Chief until her retirement. This put her at the epicenter of medical research and care quality discussions and provided a viewpoint that few others share.

As I proposed above about Bill Gates, when Marcia Angell says something, pay attention.

Where do you find 'more than one' study showing roughly the same results? Cochrane!

Cochrane publishes meta-studies or systematic reviews. These summarize multiple individual studies of the same issue and evaluate both the study methodologies and outcome results.

Here's a simple graphic description to explain. We'll call the study shown above of 2 groups using Drug X the Massachusetts study. Below is a systematic review summary of it and several other studies:

Study Name	# Heart Attacks Prevented / 100 people	# Years Studied
Massachusetts study	3	5 years
Connecticut study	4	4 years
Wisconsin study	4	3 years
Minnesota study	5	4 years

You see a trend. Cochrane reviews not only each study's outcomes but also each study's methodology. In this case, we have 4 high quality studies meaning the researchers assembled the study and interpreted the data correctly. With each additional high quality study showing roughly the same result, we gain more confidence in the conclusion.

But now add the East German study.

Study Name	# Heart Attacks Prevented / 100 people	# Years Studied
Massachusetts study	3	5 years
Connecticut study	4	4 years
Wisconsin study	4	3 years
Minnesota study	5	4 years
East German study	14	2 years

Whoa. In this clearly absurd hypothetical, the researchers discovered that the East German study was only of former female Olympic athletes over age 85 who took performance enhancing drugs for at least 10 years. Not a very representative population.

Cochrane calls those results 'low quality' and writes its summary of the other four studies, concluding that Drug X prevents about 4 heart attacks in 4 years.

Most medical researchers see systematic reviews as the highest quality medical evidence and Cochrane as an excellent source of this information, if not the best.

Cochrane groups its reviews by topic, including

> Pregnancy and childbirth – 617 reviews

> Heart and circulation – 680 reviews

> Cancer – 660 reviews, and so on.

Cochrane summaries are short, free, often around 400 words, generally written in plain English and all containing a short Author's Conclusion.

Here are screen shots of couple conclusions for illustration purposes to show how easy these are to understand.

Joint lavage for osteoarthritis of the knee. I don't know exactly what joint lavage is but I understand the author's conclusion:

Authors' conclusions:

Joint lavage does not result in a relevant benefit for patients with knee osteoarthritis in terms of pain relief or improvement of function.

Pretty clear. It doesn't reduce patient knee pain.

Second, beta blockers for hypertension. Here Cochrane used a couple of industry abbreviations, RCT for Random Controlled Trial, the type of comparative study we've been discussing, and CVD for Cardio-Vascular Disease. Their reviews are *pretty* easy to read, not *completely* easy! The underlines are hot links on their website.

Authors' conclusions:

Most outcome RCTs on beta-blockers as initial therapy for hypertension have high risk of bias. Atenolol was the beta-blocker most used. Current evidence suggests that initiating treatment of hypertension with beta-blockers leads to modest CVD reductions and little or no effects on mortality. These beta-blocker effects are inferior to those of other antihypertensive drugs. Further research should be of high quality and should explore whether there are differences between different subtypes of beta-blockers or whether beta-blockers have differential effects on younger and older people.

You get the idea. People with hypertension – high blood pressure – who take beta blockers can expect modest cardiovascular disease reduction but no mortality benefit. These medications provide less patient benefit than other antihypertensive drugs.

Cochrane offers hundreds of study summaries like these that tell *how well* various medical interventions work from an unbiased, methodologically sophisticated orientation.

How would a patient use this information? I'll present a personal example of my sore left knee as a case study.

I scheduled an appointment with my primary care physician, then looked up 'knee pain' on Cochrane. I found lots of studies including

- Osteotomy for knee osteoarthritis
- Arthroscopic debridement for osteoarthritis of the knee
- Knee braces, sleeves or straps for treating anterior knee pain
- Joint lavage for osteoarthritis of the knee
- Transcutaneous electrostimulation for osteoarthritis of the knee
- Braces and orthoses for osteoarthritis of the knee
- Therapeutic ultrasound for osteoarthritis
- Joint corticosteroid injection for knee osteoarthritis

Unfortunately, I don't know what some of those interventions entail.

So I went to the Author's Conclusion of each and summarized them on a 1-page memo to my doctor that I delivered to his office the day before my appointment. Here's the relevant bit reproduced here verbatim:

My research: Most non-surgical interventions don't benefit patients much if at all. Summaries from Cochrane.org, best website I know:

- Arthroscopic debridement = no better than placebo
- Joint lavage = no better than placebo
- Osteotomy = possibly appropriate for medical compartmental osteoarthritic, low risk, but limited comparative study data
- Transcutaneous electrostimulation = results no better than placebo
- Braces and orthoses = data inconclusive but probably little to no pain reduction benefit
- Doxycycline = minimal to non-existent pain reduction benefit
- Corticosteroid injections = possible short term benefit but little evidence of benefit after 6 months
- Exercise program / physical therapy = probably somewhat effective, about same as non-steroidal anti-inflammatories
- Non-steroidal anti-inflammatories = slightly more pain reduction than placebo but probably only short term (based on studies of anti-inflammatory impact on back pain and tennis elbow)

His opening comment when we met: Good research. It helped him move from Atul Gawande's 'informative' position to 'interpretive', saved us together a lot of time ('it doesn't work, move on') and me potentially a lot of money.

Preparing the memo took me about 15 minutes. Why wouldn't a patient do this?

Reading the memo took my doctor less than a minute. Again, why wouldn't a patient share this information with his/her physician prior to meeting? It helps your doctor prepare. Gawande seems to think it's a good idea; my doctor agreed.

The obvious caution: 1-page memo means **<u>one page</u>** (!) not a rambling treatise on your overall medical condition. Have some compassion for your doctor.

Now let's return to atenolol, the beta blocker discussed briefly above. It's moderately widely prescribed; statistica.com estimated 30 million prescriptions in 2016.[16]

Cochrane's entire write up was 420 words. Here's a screen shot of the beginning to illustrate readability and presentation.

What is the aim of this review?

The aim of this Cochrane Review was to assess whether beta-blockers decrease the number of deaths, strokes, and heart attacks associated with high blood pressure in adults. We collected and analysed all relevant studies to answer this question and found 13 relevant studies.

Are beta-blockers as good as other medicines when used for treatment of adults with high blood pressure?

Beta-blockers were not as good at preventing the number of deaths, strokes, and heart attacks as other classes of medicines such as diuretics, calcium-channel blockers, and renin-angiotensin system inhibitors. Most of these findings come from one type of beta-blocker called atenolol. However, beta-blockers are a diverse group of medicines with different properties, and we need more well-conducted research in this area.

The review then discussed the methodology and main findings. Again, a screen shot as I don't want to misrepresent their presentation.

We found 13 studies from high-income countries, mainly Western Europe and North America. In the studies, the people receiving beta-blockers were compared to people who received no treatment or other medicines. The studies showed the following.

Beta-blockers probably make little or no difference in the number of deaths among people on treatment for high blood pressure. This effect appears to be similar to that of diuretics and renin-angiotensin system inhibitors, but beta-blockers are probably not as good at preventing deaths from high blood pressure as calcium-channel blockers.

Beta-blockers may reduce the number of strokes, an effect which appears to be similar to that of diuretics. However, beta-blockers may not be as good at preventing strokes as renin-angiotensin system inhibitors or calcium-channel blockers.

Beta-blockers may make little or no difference to the number of heart attacks among people with high blood pressure. The evidence suggests that this effect may not be different from that of diuretics, renin-angiotensin system inhibitors, or calcium-channel blockers. However, among people aged 65 years and older, the evidence suggests that beta-blockers may not be as good at reducing heart attacks as diuretics.

You learn pretty quickly that beta-blockers don't do a very good job of preventing heart attacks or death.

But here's the twist to show why I think Cochrane is such an important tool for patients: in September 2018, a year after publication of this Cochran review, the US experienced a nationwide shortage of atenolol with 'increased patient demand' a reason according to the American Society of Health-System Pharmacists' announcement. [17]

Increased demand for a blood pressure lowering medicine that doesn't prevent heart attacks or deaths and works less well than alternatives? Interesting. Highlights, to me at least, the need for more patient education about medical care outcomes.

The knock on Cochrane is that it might be slightly out-of-date. After all, it takes 5 years to complete a 5-year study and longer to amass enough studies to warrant a systematic review. In my opinion, though, this is a minor, relatively rare problem since so many new medications and treatments are really just modifications of previous ones. Be sure to discuss this issue with your doctor.

I'm comfortable relying on Cochrane for my own medical information and my family's.

ChoosinglyWisely

I use ChoosingWisely second, after Cochrane. Its website, ChoosingWisely.org, lists tests and procedures that patients should question and perhaps avoid. These are often shown to be beneficial in *studies* but overused in the *real world*.

ChoosingWisely is funded by the American Board of Internal Medicine Foundation, Consumer Reports and the Robert Wood Johnson Foundation. Its lists of treatments to avoid come from the physician specialty organizations whose members prescribe and perform those tests and procedures.

In other words, organized medicine, the medical establishment if you will, here advises patients not to get the treatments it provides and profits from. Revolutionary!

Some 80+ medical specialty organizations participate, like the American Academy of Dermatology, American College of Cardiology and American Academy of Family Physicians. The boards of directors of each generally approve the list of interventions to question or avoid.

Let's use cardiac stress tests as a case study to see the potential power and impact of ChoosingWisely.

Americans under 65 years old get about 2 million stress tests annually averaging around $4342 each for an $8.7 billion US market. [18] Plus Medicare beneficiaries.

Here are stress test recommendations from various ChoosingWisely partners, a non-exhaustive list but enough to give the general idea:

- Don't perform stress cardiac imaging or advanced non-invasive imaging in the initial evaluation of patients without cardiac symptoms unless high-risk markers are present *from the American College of Cardiology*

- Don't perform annual stress cardiac imaging or advanced non-invasive imaging as part of routine follow-up in asymptomatic patients *from the American College of Cardiology*

- Avoid cardiovascular testing for patients undergoing low-risk surgery *from the Society for Vascular Medicine*

- Don't perform stress cardiac imaging or coronary angiography in patients without cardiac symptoms unless high risk markers are present *from the American Society of Nuclear Cardiology*

- And several more

ChoosingWisely also explains why. Again, the American College of Cardiology:

> Performing stress cardiac imaging or advanced non-invasive imaging in patients without symptoms on a serial or scheduled pattern (e.g., every one to two years or at a heart procedure anniversary) rarely results in any meaningful change in patient management.

> This practice may, in fact, lead to unnecessary invasive procedures and excess radiation exposure without any proven impact on patients' outcomes.

Important point here: unnecessary and overused tests and procedures are not always benign. They represent potential patient harms by, for example, identifying harmless abnormalities. (I'll say much more about this in Chapter 3.)

But once identified, doctors may be unsure how much risk each abnormality poses and patients often want treatment, sometimes asap. That's the 'unnecessary invasive procedures' comment above. These invasive procedures pose risks including infection and surgical error plus other patient harms like time off from work and family impacts.

We're still in the early stages of learning how frequently all this happens.

Cardiac stress testing is just 1 of dozens of ChoosingWisely topics and examples. I could have used others ranging from imaging for back pain to imaging for uncomplicated headaches, or antibiotic prescriptions for ear infections to population based Vitamin D screening or more. Click around ChoosingWisely for a few minutes to see the topic range.

But I want to move on to show the potential impact on patients and payers. We'll use the State of Washington's 'First Do No Harm' report that applied ChoosingWisely's analysis to 2.4 million commercially insured patients in 2018 for our estimates.

This report was prepared by the Washington Health Alliance, Washington State Medical Association and Washington State Hospital Association, i.e. pretty much everyone who's involved in organized medicine in Washington.

'First Do No Harm' analyzed 47 commonly overused medical tests and procedures through the ChoosingWisely lens, things like cardiac stress tests, and found little to no patient benefit 45% of the time. About 1.3 million commercially insured (i.e. not Medicare or Medicaid) Washington residents received at least 1 of these tests or procedures during the study's year. About 1/3 of the spending on them was wasted totaling $282 million.

Three quick rules-of-thumb for estimating total national cost impacts from Washington's data:

- Washington has about 2% of the total US population, so $282 million wasted there equals about $14 billion wasted nationally.

- Medicare and Medicaid, neither included in this study, represent about a third of all US healthcare spending, so $14 billion in commercial health insurance waste equals about $21 billion in total national healthcare system waste.

- The actual test or procedure may only represent tip-of-the-iceberg costs. That $4342 unnecessary stress test may lead to $10,000 or more in follow up tests, observations and possibly procedures.

For very rough budgeting purposes – *very* rough - I'll estimate the follow up investigations and procedures cost twice the initial unnecessary care cost. That's really just a guess on my part; the actual number may be much higher.

$282 million in Washington State's unnecessary care now becomes about $40 billion total nationally in unnecessary care plus related follow up costs. The follow up costs are also unnecessary; patients wouldn't have incurred them absent the initial unnecessary test.

Let's return to cardiac stress tests. 'First Do No Harm' estimated that 40,000 commercially insured residents have them annually, about 1/5 are wasteful representing $33 million in potential annual state savings from just this test or, using my rules of thumb above, around $4.5 billion nationally.

Here are a couple other waste estimates from the same report:

- Imaging for eye disease: 105,000 commercially insured residents had them, 74% wasteful representing $34 million in potential annual Washington state savings or another $4.5 billion nationally,

- Pre-operative lab tests: 108,000 people had them prior to low risk surgery, 85% wasteful representing $86 million in potential annual state savings or around $11 billion nationally,

- Antibiotics for acute upper respiratory and ear infections: 75,000 people had them, 98% wasteful representing $2.3 million in potential annual state savings or $300 million nationally.

I hope this brief introduction helps you appreciate ChoosingWisely's potential impact. Pretty impressive in my opinion. I encourage patients to check with ChoosingWisely prior to having medical tests or procedures and, if ChoosingWisely says 'caution', discuss this with your doctor.

Beware the knock on ChoosingWisely. It doesn't address some of the most lucrative drivers of healthcare system waste, like arthroscopic surgery for knee osteoarthritis, beta blockers to prevent heart attacks, spinal fusion surgery - perhaps a $50 billion market by itself – C-section deliveries, mastectomies, preventive percutaneous coronary intervention (PCI or angioplasty) - perhaps a $15 billion market - or lots of others. It's not perfect.

But it's useful. I hope your takeaway is that ChoosingWisely is a good resource and one that wise patients should avail themselves of.

The US Preventive Services Task Force

I recommend the US Preventive Services Task Force as the authoritative source for preventive care advice.

The USPSTF is an independent, volunteer panel of national experts in disease prevention and evidence-based medicine that makes evidence-based recommendations about clinical preventive services. It is widely regarded as the 'gold standard' in evidence-based guidelines and has been so for years.[19]

The Task Force reviews research on clinical preventive services, provides recommendations and periodically updates its work as new research becomes available. Its members are mainly practicing primary care physicians.

The Task Force uses a remarkably transparent recommendation process. First it posts draft recommendation statements online and invites public comment. Second it invites public comment on its draft research plans. Third it invites public comment on its draft evidence reviews. Finally, based both on the best current scientific evidence and all this public input, it issues letter-graded recommendations, summarized below.[20]

> **A** means **recommended.** There is a high certainty that the net patient benefit is substantial.

> **B** means **recommended.** There is a moderate certainty that the net patient benefit is moderate-to-substantial.

> **C** means **offer selectively**. There is a moderate certainty that the net patient benefit is small.

> **D** means **don't offer**. There is no net patient benefit or there is net patient harm.

> **I** means there is **insufficient evidence** available to make a recommendation.

The Affordable Care Act used these grade to determine which preventive services are free to patients at point-of-service, the A and B ones.

The USPSTF website offers short, relatively easy-to-read-and-understand explanations of its reasoning behind each grade for each service. Here, for example is a cut-and-pasted screen shot of the recommendation about Vitamin D and Calcium supplements to prevent falls.

Final Recommendation Statement
Vitamin D, Calcium, or Combined Supplementation for the Primary Prevention of Fractures in Community-Dwelling Adults: Preventive Medication

Recommendations made by the USPSTF are independent of the U.S. government. They should not be construed as an official position of the Agency for Healthcare Research and Quality or the U.S. Department of Health and Human Services.

Recommendation Summary

Population	Recommendation	Grade (What's This?)
Men and premenopausal women	The USPSTF concludes that the current evidence is insufficient to assess the balance of the benefits and harms of vitamin D and calcium supplementation, alone or combined, for the primary prevention of fractures in men and premenopausal women.	**I**
Postmenopausal women	The USPSTF concludes that the current evidence is insufficient to assess the balance of the benefits and harms of daily supplementation with doses greater than 400 IU of vitamin D and greater than 1000 mg of calcium for the primary prevention of fractures in community-dwelling, postmenopausal women.	**I**
Postmenopausal women	The USPSTF recommends against daily supplementation with 400 IU or less of vitamin D and 1000 mg or less of calcium for the primary prevention of fractures in community-dwelling, postmenopausal women.	**D**

As you can see, the Task Force sometimes gives different grades to different population groups for the same service.

Here's part of their Vitamin D and Calcium supplement grade justification:

> The USPSTF found adequate evidence that supplementation with vitamin D and calcium increases the incidence of kidney stones...For every 273 women who received supplementation over a 7-year follow-up period, 1 woman was diagnosed with a urinary tract stone...

> The USPSTF found that vitamin D supplementation does not reduce the number of falls or the number of persons who experience a fall.[21]

You can review the USPSTF recommendations relevant for you in very few minutes. It's a pretty easy-to-use site.

Wise patients can discuss the recommendations with their doctors in different ways. Some can reasonably say 'I want all the A and B services that apply to me.'

Others can reasonably say 'I only want the A services that apply to me.'

Still others can reasonably say 'I want all the A services and some of the B services that apply to me plus one specific C service.'

Contrast these with the too often 'Doc, I saw this test on TV. It's not very expensive and I want it.'

Let's compare a TV promoted service recommendation from Life Line Screening with the USPSTF's Abdominal Aortic Aneurysm Screening recommendation. The Life Line information came from http://www.lifelinescreening.com/ on October 19, 2018.

Life Line recommends abdominal aortic aneurysm screening for people over age 50 or people over age 40 with risk factors such as immediate family history of AAA. Most people, they say, will need AAA screening every three years but some, depending on risk factors and earlier screenings, may need it annually. The test costs $149 by itself or $181 as part of a 5-test package.

Now see the USPSTF's recommendation for the same AAA screening, screen shot below. It's the June, 2014 recommendation, in effect in October 2018 when I wrote this section.

Recommendation Summary

Population	Recommendation	Grade (What's This?)
Men Ages 65 to 75 Years who Have Ever Smoked	The USPSTF recommends one-time screening for abdominal aortic aneurysm (AAA) with ultrasonography in men ages 65 to 75 years who have ever smoked.	B
Men Ages 65 to 75 Years who Have Never Smoked	The USPSTF recommends that clinicians selectively offer screening for AAA in men ages 65 to 75 years who have never smoked rather than routinely screening all men in this group.	C
Women Ages 65 to 75 Years who Have Ever Smoked	The USPSTF concludes that the current evidence is insufficient to assess the balance of benefits and harms of screening for AAA in women ages 65 to 75 years who have ever smoked.	I
Women Who Have Never Smoked	The USPSTF recommends against routine screening for AAA in women who have never smoked.	D

The USPSTF says inappropriate for female non-smokers over age 65 unclear for female smokers over 65 and appropriate only for male smokers over age 65, once; Life Line says differently.

You, the patient, get to choose.

I want to point out one often-overlooked potential problem with deviating from the USPSTF recommendations or rejecting them outright: it can affect the doctor-patient relationship.

Physicians, highly trained to do right by patients, can get frustrated with patients who ignore high quality information in favor of 'I saw it on TV' or 'My cousin had it'. I worry about this since research suggests that higher levels of physician empathy lead to better patient outcomes [22] and to higher patient satisfaction with care. [23]

Shared decision making based on well informed patients using high quality information, yes.

Rejection of physician recommendations by poorly informed patients, no.

I'll wrap up this section with advice from doctors Vinay Prasad and Adam Cifu in their brilliant book Ending Medical Reversal: Never get screened more aggressively than the US Preventive Services Task Force recommends.

Consider it when next you need preventive services.

And don't forget to read Chapter 3 of this book too!

Summary and Key Take-Aways

I start with Cochrane. That tells how well a treatment works in study conditions. I always look for tests, medications and procedures that outperform the placebo by a reasonable amount. Each patient can decide what 'a reasonable amount' is for him or herself.

I then check with ChoosingWisely to see it it's overused in real life.

Then, if it's preventive care, I check with the US Preventive Services Task Force. I only consider the A graded services but that's a personal opinion. Reasonable patients can make different and equally well-informed decisions about this.

Finally, I discuss everything with my Primary Care Physician because I value his experience and wisdom. (I'd do the same with specialists but I don't use any currently.) I would no sooner make a decision without my doctor's input than I would turf all my medical decisions to him.

That's it. Preparing wisely and completely for your doctor's appointment is far simpler than many people think. Use the three sources introduced here, read the information they present thoughtfully and share your research and observations with your doctors.

I think you'll find yourself engaging in better discussions with your treaters and making wiser, more well-informed decisions as a result.

Chapter 2: Sick Visits
What to discuss with your doctor

This chapter addresses sick visits. The discussion here starts after you identify and describe a medical ailment to your doctor and he or she makes a treatment recommendation.

Chapter 3 addresses well visits.

Well informed patients focus on three issues, stated here as questions:

- How likely am I to benefit and be harmed from this medical intervention? This question includes two parts

   o First, 'what do medical studies say about this intervention?' and

   o Second, 'Is the intervention overused in the real world?' making me perhaps less likely to benefit from it than the study indicates.

- What are my alternatives? One sometimes overlooked consideration: Is doing nothing a reasonable option?

- Which hospital and specialist gets the best outcomes for my preferred treatment choice?

Patients who ask and then learn the answers tend to enjoy better outcomes with less risk and at lower costs.

Unfortunately, most patients don't probe and few, apparently, understand the importance of these questions. Only 12% of Americans are 'medically literate' according to the US Department of Health and Human services, defined in part as understanding the likelihood of benefiting and being harmed by a given medical intervention.[24]

Low medical literacy according to HHS, is linked to poorer health outcomes, higher hospitalization rates and higher healthcare costs.

'Illiterate' doesn't mean stupid! Instead it means 'hasn't been taught.' I, for example, am illiterate in several hundred things including Italian and nuclear physics. Consider this chapter a crash course in medical literacy.

The approach introduced here applies to all patients from the least sophisticated to the most, including medical professionals, several of whom have commented privately that 'I hadn't thought of that', 'useful approach' and 'can you send me that list of questions?'

Four questions that *every* patient should ask *every* clinician about *every* medical intervention: the foundation of medical literacy

- Out of 100 people like me, how many benefit and are harmed by your recommended intervention according to studies?

- Is it overused in the real world?

- Would most physicians make the same recommendation or might some suggest something different?

- How many patients like me do you, the specialist you're referring me to and the hospital you use, treat annually?

Question 1: Out of 100 people like me, how many benefit and are harmed by your recommended intervention according to studies?

We'll discuss below the key components of this question:

- Out of 100
- Like me
- Studies
- Benefit
- Harm

Cochrane is the best easily-accessible, public source for answers. Maybe review that write up in Chapter 1 after you read this section.

Ask **'out of 100'** to get a number for your answer. '5' or '39' convey more information than 'few' or 'many'.

The number tells the likelihood that *you* will benefit. It comes from the comparative studies we discussed in Chapter 1 and is a statement of fact.

'Few', 'very few', 'quite a few' or 'many' act more like value judgements and mean different things to different people. Is 5 'very few' suggesting that your doctor doesn't much like it or 'quite a few' suggesting that your doctor seems to favor it? We have no standard definitions, but these phrases can impact patient decisions.

- Patient: 'I wish you had told me it was only 5.'
- Doctor: '5 is quite good.'

Inappropriate answers to 'out of 100' include

- '50% better than'. 50% better *than what*? Percentage answers are a red flag indicating something other than scientific clarity. The only thing you learn from '50% better than' is that something is 50% worse, which may confuse more than it illuminates.

Consider the lottery ticket example. I recently bought 2 tickets using my standard lucky numbers that have never won, but that's a different story.

Then, since the payoff was over $1 billion, I bought a third. That increased my chance of winning by 50%, taking it from 2 in 50 million to 3 in 50 million.

o 50% sounds big.

o 1 in 50 million is small.

o Beware percentage answers!

- 'reduces your risk by 36%'. Same issue. Unless you know your initial risk, you can't determine what a 36% reduction means.

- 'the guidelines say'. Another red flag. You asked for a number and got a bunch of words.

  Here's the skinny on medical guidelines. Some 300 organizations promulgate 2300 different guidelines. Some are good but others, according to Dr. Otis Brawley, Chief Scientific Officer of the American Cancer Society until 2018, are commercial documents developed by some organization or other that wants to profit from the recommendations.[25]

  Guidelines don't tell *your* likelihood of benefiting from a particular medical intervention, only that some group thinks it's a good idea for an oft undisclosed or opaque reason. The guideline may benefit them financially, you medically or both – you don't know!

Get a number as the answer to your question to eliminate this confusion.

Ask about **'people like me'** since treatments can have different outcomes for different demographic groups, 25 year old male athletes or 85 year old female diabetic smokers. As one example, in 2013 the FDA halved the recommended initial dose of zolpidem, a sleep medication, for women after discovering that women had more in their systems the following morning than men, putting them at higher risk of impaired driving.[26]

'Studies' mean the comparative studies introduced in Chapter 1. Here's that graphic again showing that Drug X prevents 3 heart attacks per 10 people over 5 years.

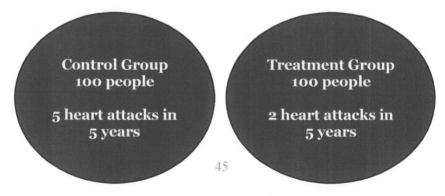

Control Group
100 people

5 heart attacks in
5 years

Treatment Group
100 people

2 heart attacks in
5 years

'Studies' specifically doesn't mean 'biological or anatomical knowledge of the human body.' **That's a bad way to predict the likely outcomes of a medical intervention!**

We often can't meaningfully isolate one or few biological components or metabolic processes and predict patient outcomes; our bodies contain too many that interact and impact each other. When we attempt to predict outcomes from one or a few stagnant components, we sometimes focus on the easiest to identify and quantify, not necessarily the most important.

We sometimes also fail to account for ongoing interactions - side effects or rebound impacts - that can modify the stagnant-state model.

Both errors can lead to patients sometimes receiving inappropriate care.

Take cholesterol lowering drugs, for example. These can lower the heart attack risk, high cholesterol being a risk factor for heart attacks.

But they can also cause muscle discomfort, cholesterol being necessary for muscle development. Muscle discomfort can lead to less exercise, itself is a risk factor for heart attacks. I'll quantify some of these impacts later in this chapter.

Researchers have identified hundreds of specific instances in which relying on biological justifications for medical interventions led in the wrong direction. I'll discuss only three briefly here but could have included many more like surgery to treat knee arthritis, stenting to prevent heart attacks, spinal fusion surgery to reduce back pain or flecainide to stabilize heart rhythm and reduce heart attacks.

You can find a good list in the Appendix to Prasad and Cifu's 2015 book Ending Medical Reversal.

First, niacin to prevent heart attacks:

Niacin, a B vitamin, has been shown in tests to raise good cholesterol (HDL) and lower bad (LDL). More good and less bad cholesterol is associated with fewer heart attacks. Based on this, Abbott Labs developed Niaspin, an extended release niacin pill that generated about $900 million in 2009 sales.[27]

In 2011 though, the AIM-High study showed that extended release niacin generated no significant reduction in cardiovascular events.[28]

Remember Marcia Angell's recommendation from Chapter 1, that wise patients should rely on more than 1 study showing basically the same thing?

In 2013 a second study on the niacin drug Tredaptive showed pretty much the same thing, no difference in coronary events.

Cochrane then weighed in with their 2017 study 'Niacin for people with or without established cardiovascular disease', concluding [i]

Moderate- to high-quality evidence suggests that niacin does not reduce mortality, cardiovascular mortality, non-cardiovascular mortality, the number of fatal or non-fatal myocardial infarctions, nor the number of fatal or non-fatal strokes but is associated with side effects. Benefits from niacin therapy in the prevention of cardiovascular disease events are unlikely.

The clinical-recommendations-based-on-biology notion didn't hold up here at all. Dr. Steven Nissen, Chief of Cardiology at the Cleveland Clinic summarized the 2013 Tredaptive findings: [29]

It raised the good cholesterol, it lowered the bad cholesterol, it didn't improve clinical outcomes.

That is a stunning finding.

---

[i] The underlined words in this screen shot are hot links.

Yes, stunning...$900 million later.

Second, ezetimibe to lower cholesterol:

> Ezetimibe, commonly sold as Zetia, blocks absorption of cholesterol, especially LDL, in the intestines, not in the liver like statins. It can be used instead of, or in addition to, statins to control cholesterol levels. Zetia generated $3+ billion in annual sales for several years in the 2010s.

> That Zetia lowers bad cholesterol is clear from various studies and their advertising.[30]

> But that doesn't necessarily translate to fewer deaths from heart attacks.

> Here's Cochrane's conclusion from its 2018 study 'Ezetimibe for the prevention of heart disease and death':

> - Moderate- to high-quality evidence suggests that ezetimibe has modest beneficial effects on the risk of CVD endpoints... ('modest' is the researcher's way of saying 'probably, a little, under certain circumstances, for some people')

> - ...but it has little or no effect on clinical fatal endpoints. (That's the researcher's way of saying 'dying'.)

> Cochrane's report also stated that

>> Ezetimibe is a non-statin drug that can reduce the blood lipids levels by inhibiting cholesterol absorption, but whether it has beneficial effects on heart disease and death remains uncertain.

> Zetia's advertising admitted this though obliquely, in ways that only highly trained patients could understand, stating *Unlike some statins, Zetia has not been shown to prevent heart disease or heart attacks.*[31]

Well trained patients understand that *'has not been shown to prevent heart disease or heart attacks'* means one of two things.

- Either it *has not* been studied, subjecting it to Prasad's Law (coming up shortly) or

- It *has* been studied but didn't generate heart disease or heart attack reduction.

Untrained patients assume this phrase is legal gobble-di-gook and doesn't mean anything about clinical outcomes.

Well trained vs. untrained patients. Tens of billions of dollars on a drug that has little to no life saving benefit.

Third, vertebroplasty, the process of injecting medical grade cement into fractured vertebra (back bones) to reduce back pain.

In 2008, the US market for vertebroplasty was around $245 million.

Then in 2009 the New England Journal of Medicine published two studies – remember Marcia Angell again - comparing vertebroplasty to a control group that received lidocaine (a topical skin numbing agent), massage and aromatherapy to reproduce operating room smells.

- The Australian study found 'no beneficial effect' of vertebroplasty compared to the control group.

- The Mayo study found the same thing, that patient improvements were similar in the placebo and experimental groups.[32]

Vertebroplasty, in other words, worked as well as, but no better than, the safer and far cheaper placebo.

The market then shrunk to around $70 million in 2016 [33] indicating that the procedure had, in fact, been overprescribed.

But even that shrunken market might be too big. Cochrane weighed in with its November, 2018 review, Vertebroplasty for Treating Spinal Fractures due to Osteoporosis, concluding

Authors' conclusions:

We found high- to moderate-quality evidence that vertebroplasty has no important benefit in terms of pain, disability, quality of life or treatment success in the treatment of acute or subacute osteoporotic vertebral fractures in routine practice when compared with a sham procedure. Results were consistent across the studies irrespective of the average duration of pain.

What about all those thousands of patients who spent hundreds of millions on it previously? They undertook surgical risks without any potential benefit – the subsequent comparative studies prove that. At best they received no benefit; at worst they were harmed.

They and their doctors had relied on biological justifications, not comparative study results, in their treatment decision making.

These examples – extended release niacin, ezetimibe and vertebroplaty – are but three of hundreds that show how medical theory can lead patients to the wrong treatment decisions.

Steven Nissen, the Cleveland Clinic cardiologist, summarized the problem: 'the road to hell is based on biological plausibility.' [34]

Note that in all three examples, and many more not listed here, *outcome based* studies showed that *theory based* interventions worked poorly, if at all, though with a time lag; Cochrane's studies sometimes came a decade or so after common use.

That raises a difficult question: How often do comparative study results contradict theory-based medical practices?

Vinay Prasad, the brilliant researcher who studies this, estimates that about half of medical theory-based treatments are proven ineffective or harmful by subsequent comparative studies.

I call this Prasad's Law in my classes:

If it hasn't been subjected to comparative studies, it's ineffective or harmful about half the time.

I told this to Dr. Prasad at a medical conference in 2018 and asked if I got it right. He responded affirmatively.

- If you rely on comparative study-based facts in your medical decision making and you'll make wise decisions.

- But if you rely on biological theory...not so much.

- And, my own suggestion, if it hasn't been studied by Cochrane, proceed cautiously.

Consider spinal fusion surgery in light of all this. I want to show how understanding Prasad's Law and its implications can potentially improve patient outcomes, reduce patient risk and save payers a great deal of money.

Americans get about 500,000 fusion surgeries annually,[35] costing perhaps $100,000 each, the total of pre-surgery, surgery and rehab.,[36] making it a $50 billion annual market.

But these surgeries have never been subjected to independently funded, high quality comparative studies.[37] Cochrane's two most recent reviews said as much.

- "To date no trials have compared spinal fusion surgery with no treatment, placebo or sham surgery." *Effectiveness of Surgery for People with Leg or Back Pain Due to Symptomatic Spinal Stenosis 2016*

- "We cannot conclude whether surgical or non surgical treatment is better." *Surgical vs. Non Surgical Treatment for Lumbar Spinal Stenosis, 2016*

This is a big red flag to wise patients who see spinal fusion surgery as untested and therefore ineffective or harmful about half the time.

That may prompt them to try other (tested) intervention first ... which, in turn, may actually result in more benefit with less risk and at far lower costs.

The *financial* savings result from wiser *clinical* decisions – not the other way around - just like in the Niaspin, vertebropasty, ezetimibe and beta blocker examples above.

Let's turn now to what '**benefit**' means in our question 'out of 100 people like me, how many benefit?' We have two basic types called 'indicators' and 'patient events.'

- Indicators, sometimes called surrogates, include things like cholesterol, blood pressure and bone density, levels we can measure. You wouldn't know about them unless you test. These *seem to suggest* something about your likelihood of having a patient event.

- Events are things that really matter to a patient like heart attacks, leg amputations and death, all pretty clear.

How closely do indicators and events correlate? How often does dangerous cholesterol, for example, correlate with heart attacks?

Lipitor, a cholesterol reducing medication, provides insights from its ads. Here's one from the Wall Street Journal, Dec 4, 2007. Lipitor sold in the billions of dollars annually in the early 2000s.

THE WALL STREET JOURNAL Tuesday, December 4, 2007 A13

We learn from the bottom left how much Lipitor impacts heart attack rates: 3% of patients taking a sugar pill or placebo had a heart attack compared to 2% of patients taking Lipitor. (Reminder from elementary school arithmetic and high school Latin: percent means 'out of 100'.)

Let's put this into our comparative study diagram.

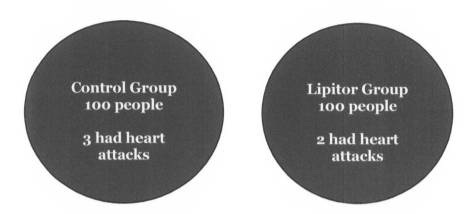

Control Group
100 people

3 had heart
attacks

Lipitor Group
100 people

2 had heart
attacks

Lipitor, according to its ad statement, prevents 1 heart attack per 100 at risk people who take it.

Lowering cholesterol with Lipitor doesn't, in other words, correlate with preventing heart attacks 99% of the time. Their numbers, not mine. I assume this represents a best case scenario. Why would Lipitor advertise anything else?

Interestingly, this ad didn't state the time period or specific population. This is important because some research suggests that statin benefits fade as people age. [38]

Lipitor's benefit estimate is in line with Cochrane's 2013 study **Statins for the primary prevention of cardiovascular disease.** Cochrane estimated that 1.8 people per 100 who took statins as primary prevention for 5 years would avoid a cardiovascular event.

Indicators other than cholesterol, of course, like blood pressure or blood sugar, may correlate better or worse with other patient events.

Be sure to define 'benefit' when you ask your doctor 'out of 100 people like me, how many benefit?' If you mean indicators, learn how closely they correlate to patient events. Cholesterol lowering or heart attack prevention? As this Lipitor ad and Cochrane summary tell us, they're not the same.

Ditto for blood pressure, blood sugar, bone density and other indicators.

Harms always exist in medical care. Comments like 'harms are infrequent and minor' are as meaningless as benefits are 'frequent and major'. Aim for actual numbers.

Let's stick with the statin example since we've just estimated the benefits, to estimate harms. Again, this is for illustration purposes only. We could do similar analyses about other indicators.

- One study suggests that around 5 in 100 people who took statins reported muscle aches.[39]
- Another reported that 2 in 100 patients reported muscle tendonitis, primarily in their Achilles tendons. [40]
- A third found that about .4 in 100 developed diabetes.[41] That's a little less than half of 1 suggesting that for every 2 ½ heart attacks prevented, 1 person develops diabetes.

Diabetes increases your chance of having a heart attack, stroke, kidney failure or limb amputation, among other things, which can mitigate the statins' long term benefit.

Now you can answer the question 'out of 100 people who take statins, how many benefit and are harmed?' with these rough estimates:

- 1 in 100 avoids a heart attack
- 5 in 100 report muscle aches
- 2 in 100 report muscle tendonitis
- .4 in 100 develop diabetes

After all that work, you are finally in position discuss all this with your doctor and decide if statins work well enough to take. Phew!

As part of that process, discuss with your doctor whether statins provide sufficient benefit to continue on them if you experience muscle discomfort. Different people can reasonably answer differently.

Let's now turn to the **time** period issue. You may be prescribed a medication at age 50 for life. Has it been tested for 35 years? Consider these cases:

- In a 6-month trial of 8000 people, the arthritis drug Celebrex showed lower rates of stomach and intestinal ulcers and related complications than two other arthritis drugs, diclofenac and ibuprofen. Some doctors and patients presumably made medication decisions based on those facts.

But the full 12-month test showed Celebrex's safety advantage disappeared.[42]

- In the 1990s after a total of 42 clinical trials, the FDA and approved several new antidepressants including Prozac, Paxil, Zoloft, Celexa, Serzone and Effexor. Patients may take these drugs for years.

But the majority of those 42 trials lasted just 6 weeks.[43]

Long term study data may be difficult to obtain. Clearly worth discussing with your doctor, but together, you and he or she will need to make some assumptions and rely on best guesses about the time period studied and the likely impact on you. Good luck on this.

One final thought about asking this question. Don't freestyle with the wording. Phrase the question exactly as written here, 'out of 100 people like me, how many benefit and are harmed' over a given time period. You may feel peculiar simply parroting these phrases, but your doctor will understand and appreciate your clarity.

At least that's the feedback I get from people who ask it!

Question 2: Is it overused in real life?

Testing sometimes shows that a treatment works well on a narrowly specified group of patients but, in the real world, doctors may offer it more widely. The wider group may have a different risk profile and consequently enjoy less benefit.

A good source of overuse information is ChoosingWisely. You may want to review that part of Chapter 1 after reading this section. More technical, less user-friendly sources also exist like Atlas of Healthcare. Your doctor may know about this

I'll suggest three main reasons for treatment overuse and discuss below:

- Physicians want to broaden use to all patients who could possibly benefit, rather than restrict use only to a narrowly defined group of patients.

- Treaters – clinicians and hospitals – develop treatment routines which may resist change over time.

- Treaters gain financially from various treatments.

Expand the patient base. Doctors may think 'studies show this medication benefits people *somewhat like* my current patient. I think my patient may also benefit. I'll prescribe it.'

This is not necessarily bad and patients may well often benefit. Deciding when to expand the patient base is part of the art of medicine.

My only point here: discuss this with your doctor so together you can decide how best to move forward.

Treatment routines and hospital standard operating procedures. Physicians and hospitals sometimes develop treatment habits and, through practice, get very good at them. These sometimes resist change.

I'll let a specific case stand for the general situation. It's a comparison of mastectomy rates among Medicare beneficiaries in Massachusetts and Connecticut over time. Though the rate of mastectomies has fallen over time in both state, Connecticut always performs more per 100,000 Medicare women.

Mastectomies per 100,000 Medicare women
Connecticut = top line
Massachusetts = bottom line
Data from the Dartmouth Atlas of Healthcare

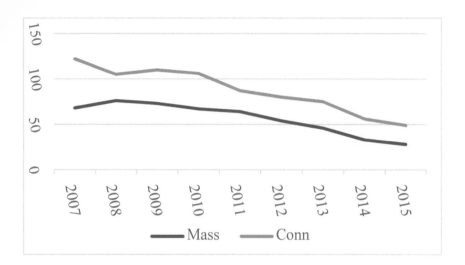

The interesting and noteworthy point here is that breast cancer incidence and mortality are the _same_ in both states, suggesting that Connecticut hospitals routinely overuse this procedure.[44] In other words, Connecticut hospitals perform more mastectomies on the same patient population as Massachusetts hospitals but don't generate better outcomes as measured by breast cancer death rates.

That's the classic definition of overuse.

Expanding on this theme, Dr. John Wennberg, founder of the Dartmouth Institute for Healthcare and the Dartmouth Atlas, estimates that the best way to predict your chance of receiving a specific treatment is to learn the rate of that treatment in your region 10 years ago. That's because physicians and hospitals often rely on routines, he suggests, not the most up-to-date evidence.

In the mastectomy example above, the chance of a Connecticut woman with early stage breast cancer receiving a mastectomy is higher than a similar woman in Massachusetts both historically and, according to Wennberg, also into the future.

Wennberg, his team and other researchers provide similar insights about back surgery, hip replacements, knee replacements, coronary angiography, radical prostatectomies, tonsillectomies, c-sections, medical prescribing and more. It's fascinating research but a pretty heavy lift to read.

My advice here, again, is simply to ask your doctor if his / her recommended treatment procedure is overused in the real world. Good, well informed physicians will provide thoughtful answers. Those with less wisdom may more likely respond 'I never do unnecessary procedures.'

Then you, the patient, get to decide how to proceed.

If you get stuck here, consider adding the next question 'would most doctors make the same recommendation or might some suggest something different?' It may help you get unstuck. We'll get to that discussion shortly.

Financial incentives. The impact of financial considerations can perhaps best be summarized with Upton Sinclair's 1935 aphorism: It is difficult to get a man to understand something when his salary depends on him not understanding it.

Though physicians routinely claim never to consider their own remuneration when recommending treatments, the evidence too often shows otherwise. For example

- One study found that a 2% increase in payment leads to a 3% increase in care provision.[45]
- Another found that cataract surgery rates dropped 45% after physicians in one practice went on salary and were no longer paid per surgery.[46]
- Yet another found that Lupron sales increased dramatically when injecting it became a billable event. [47]

You get the idea. Remuneration can impact medical advice.

This problem is compounded when insurance covers an intervention. That makes it appear free to the patient. Less wise patients may think 'it's free to me so I might as well have it.' The too often result: the patient wants it because it's free, the physician or hospital benefits financially from it and bingo, we get overuse.

The impersonal question 'is it overused in real life?' removes a specific physician's reimbursement from consideration. When combined with our next question, it goes a long way to protecting you from receiving inappropriate care.

One final thought about discussing overuse with your doctor. Patients often praise their doctor as wonderful, trustworthy, insightful, caring, compassionate etc., with phrases like 'he/she really understands me' and 'I really trust his/her advice.' Discussing overuse with your doctor may deepen those sentiments and lead to an even better relationship.

Or not.

Question 3: Would most doctors make the same treatment recommendation or might some suggest something different?

This question expands your range of treatment choices. You have options about 85% of the time according to Dartmouth's Wennberg.

- Surgery or physical therapy?
- Physical therapy or medications?
- Mastectomy or lumpectomy?
- More aggressive surgery or less?
- Lifestyle change or medication?
- Treat now or watch and wait?

Outcomes are often similar but the process, time, recovery, pain, cost and personal impacts may vary significantly.

Patients who ask this question and explore their options switch to choose less invasive, less risky care about 30% of the time according to studies.

The underlying notion here is called 'preference sensitive care', meaning similar patients with the same medical situation can choose different care and all be right: The right care for you may be wrong for a someone else who's medically very much like you.

I suggest patients try to quantify and compare the benefits and risks over three phases of care: the initial treatment, the short term and the long term. Try to complete this chart, with your doctor's help if possible. You can define benefits, harms, short term, medium term and long term as you feel appropriate.

	Option A	Option B
Benefits and harms of intervention		
Benefits and harms over the short term		
Benefits and harms over the medium to long term		

I'll offer as an example, some questions to compare a hypothetical surgery to physical therapy. This is a non-exhaustive list and intended only to stimulate your thinking.

Each patient can define short, medium and long term for themselves; there's no standard industry definition.

	Surgery	Physical Therapy
Benefits and harms of intervention	How likely is an infection or surgical complication? How long will I be hospitalized? How much pain, for how long? How long until I return to work? Will I need home assistance?	How many sessions will I need? How much pain? When will I know if the therapy is working?
Benefits and harms over the short term	How long to regain my strength and range of motion? How often are patients readmitted to the hospital? How often do	How often do patients quit physical therapy and opt for surgery? How many patients are satisfied at 6 and 12 months?

patients report pain
reduction at 6 and 12
months?
What's involved in
rehab?

Benefits and harms over the medium to long term	How often do patients require a 2nd surgery? How many are satisfied at 48 months?	How often do patients op for surgery within 48 months? How many patients are satisfied at 48 months?

You choose the issues that concern you the most. As you work through this process and discuss it with your doctor, you may see that you and he/she have different concerns and questions. Why?

Advice givers – your doctor in this case – may have different orientations from advice receivers – you. Though you both clearly want successful treatment outcomes, you don't necessarily share all the other considerations. Here's a partial list for example.

Advice giver (doc)
- Success
- Professional norms, legal issues
- Insurance carrier regs and practices
- Potential guilt
- Treatment orientation and experience
- Compensation

Advice receiver (patient)
- Success
- Amount of pain
- Time away from work
- Family impact
- Self image post treatment
- Out of pocket costs

A quick personal comment about advisor guilt.

My best friend from college, a Briton, asked my advice about PSA screening when it became commonly available there some years after here. I don't think it's a good test and didn't have it myself.

But I hesitated to advise him against, in part because I'm not his doctor, in part because I didn't want my biases to influence his decisions and also in part because I wanted to avoid potential guilt if - horrors – he later became diagnosed with untreatable, late stage prostate cancer as a result of skipping the test.

That experience gave me a small glimpse into physician responsibilities, the importance of avoiding future guilt and the default to 'sure, go ahead and have the test' advice. It really had an impact on me.

It also showed how advice givers can downplay treatment risks: I worried more about my biased advice possibly resulting in death rather than in his impotence or incontinence, two known side effects of prostate removal surgery...but that might just reflect my own biases!

Related to this, there are sometimes also wide gaps between what patients want and what doctors think patients want.

One survey, for example, found that doctors think 71% of patients with breast cancer rate keeping their breasts as a top priority while only 7% of patients agreed; something else was presumably more important to many of them.

Another found that doctors believe 96% of breast cancer patients considering chemotherapy rate living as long as possible as the top priority while only 59% of patients agree.[48]

The chart and discussion above show why asking 'what would you do if you were me, Doc?' is an unfair question; your doctor *isn't* and *cannot be* you.

Let me summarize this section on treatment choices with three comments from thought leaders in this field. First, from the Dartmouth Atlas:

When patients are fully informed about their options, they often choose very differently from their physicians.

Second, from Laura Landro, health columnist for the Wall Street Journal:

When patients understand their choices and share in decision making with their doctors, they tend to choose less invasive and less expensive treatment than they would otherwise have received. [49]

Third from Michael Porter, Harvard Business School's great strategy professor and Elizabeth Teisberg of the University of Virginia's Darden School of Business in their massive study Redefining Health Care:

Patients who share in decision making often choose more conservative, less expensive treatments and less surgery...experience better outcomes and lower costs...by obtaining care consistent with their values and preferences and through selecting only care consistent with medical evidence. [50]

I hope this discussion and the various charts and questions introduced will help you choose wisely among your treatment options.

Once you decide which treatment you prefer, *and only then*, move to Question 4, the specialist and hospital choice question.

Question 4: How many patients like me do you treat annually?

Provider experience is a good proxy for specialist and hospital quality. It's not perfect but rather 'pretty good', about the best available in the real world.

By and large, the greater the care provider's experience, the better the patient's outcomes. Less experience correlates with more errors, more surgical re-dos, longer hospital stays, more risk and higher mortality rates.

As a result, less experience also correlates with higher costs.

Experience seems to trump technology, medical school affiliation, fame or any other individual factor in predicting positive patient outcomes.

Consider a 2018 study of 1 million hospitalizations managed by more than 30,000 physicians that found no difference in patients' risks of death or hospital readmission according to where their doctor's medical school fell in the U.S. News & World Report medical school rankings. [51]

Graduates of famous schools, in other words, don't generate better patient outcomes than grads of less prestigious places.

Instead physician and hospital experience treating patients with your medical condition does. This chart showing thyroidectomy error rates by surgeon annual volume is a specific example of the general principle. [52]

Number operations / surgeon / year	Rate of surgical complications
1	87%
2 - 5	68%
6 - 10	42%
11 - 15	22%
16 - 20	10%
21 - 25	3%
>25	3%

Note the obvious caveat: a physician may become expert at a minimally or non-beneficial procedure, vertebroplasty, for example, discussed above, or preventive angioplasty. I strongly encourage patients to ask this question *only after* they've determined that they will likely benefit from the specific intervention.

Why does practice make perfect in medicine? Two fundamental reasons.

First, more experience allows a surgeon to identify a patient who is 'out of bounds' more quickly and to deal with that issue more promptly. How much bleeding, for example, is appropriate during a thyroidectomy? Only through experience can a physician realize quickly which patient is bleeding excessively and react promptly.

Atul Gawande – remember him from Chapter 1? - described this process in his book Complications, edited here for length and readability: [53]

The keys to perfection are routinization and repetition.

> With repetition, a lot of mental functioning becomes automatic and effortless, as when you drive a car.

He described observing a very experienced surgeon operate:

> The surgeon performed each step without pause, almost absently...he slowed down at a key point for a moment to check meticulously on the progress.

Second, staffs in hospitals with more experience treating similar patients develop the same insights and skills. Plus hospitals performing large numbers of specific procedures may have equipment or communications protocols adapted for those patients. That's why you want to use a hospital that treats lots of patients like you.

Return now to the last line of the thyroidectomy chart above. It suggests that patient outcomes don't improve once surgeons exceed about 23 procedures per year.

This is called a surgical threshold. Below it, patient outcomes are worse but above it, patient outcomes don't improve.

There's no patient advantage, in other words, to choosing a surgeon who performs 75 thyroidectomies annually instead of one performing 23. There is, though, the obvious advantage of choosing one doing 23 procedures over one doing only 5; the threshold tells us.

Researchers have developed both surgeon and hospital thresholds for various procedures. Here's one set published in 2015 for 10 common procedures: [54]

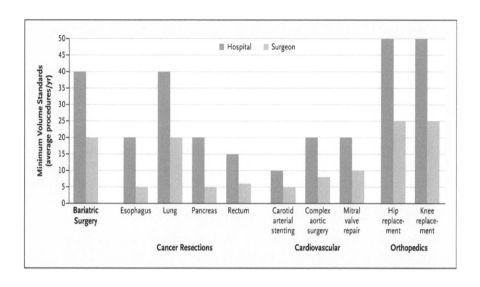

I used this information recently when a family member needed surgery and invited me to meet the surgeon during a pre-op appointment.

I declined.

I figured I might like the surgeon, which could affect my judgement and ability to help my family member decide how to proceed. Or not like him, which could also affect that ability.

Neither, though, would have any bearing on the likely surgical outcome.

Instead I was only interested in learning two things: did the hospital exceed the recommended threshold above, and did the surgeon.

Both did and the surgery went flawlessly.

Threshold estimates get updated regularly as researchers learn more and more. I advise patients to ask their doctors about current threshold recommendations as well as their own exact experience. I did this with my family member's surgeon when I finally met him, and said that I used the threshold estimates above.

He responded that my estimates were low in his opinion and that current research suggests higher numbers. Good enough. He's fully immersed in his field while I only do general research. I was happy to go with his estimates.

Feel free to press your own doctors on all this. If they don't understand your question or demean it, think twice about proceeding with them. They really should know this stuff cold.

One final note on thresholds. Some researchers affiliated with the Dartmouth-Hitchcock Medical Center, Johns Hopkins and University of Michigan Health System, notably Drs. John Birkmeyer and Peter Pronovost, have initiated a 'Take the Volume Pledge' movement. This challenges health systems to restrict procedures to the hospitals and surgeons who exceed recommended thresholds.

Birkmeyer estimates that about half the hospitals that currently perform certain procedures would stop doing them as a result. That strikes me as a big deal.

The complete phrasing of Question 4 now becomes

- How many patients like me do you treat annually? And
- Is it above the recommended threshold?

Integrating these with Questions 3 'Would most physicians make the same treatment recommendation or might some suggest something different?' helps us answer two of the most common patient questions:

- Who's the best doctor? and
- Which is the best hospital?

The answer in both cases: the ones that meet the recommended thresholds for your preferred treatment option. Feel free to get *advice* from 'the doctor that all the doctors go to' the head of the department or your cousin's doctor 'the nicest guy around.' But once you decide which treatment you prefer, choose your treaters - physician and hospital - based on their experience meeting thresholds.

Gawande made his own family member decisions this way. His then-infant son Walker who inexplicably went into congestive heart failure at 11 days old, needed a pediatric cardiologist to follow his development. A very attentive, highly competent cardiology fellow offered to follow Walker. He had put in the most time caring for little Walker, knew the case best, made the initial diagnosis, stabilized him, coordinated with the surgeons and visited Walker every day. In many ways, the best fit.

Gawande instead choose a more experienced doctor who was less involved in this case. "I was not torn about the decision," Gawande wrote, "This was my child. Given a choice, I will always choose the best care I can for him." [55] Probably good advice for you too.

A closing thought

Becoming medically literate isn't terribly difficult. Virtually anyone who puts in the effort can be successful. You just have to learn what to do or ask, and why.

The important question for patients isn't 'Can I learn to do this?' but instead 'Who can teach me?' and 'What books, manuals or classes exist to help me?'

I hope you can see how this chapter and book can help.

As you go through this process, remember that you have a partner readily available: your doctor(s), starting with your Primary Care Physician. Most docs, in my experience, welcome well-informed patient questions and discussion.

Most, in other words, want to be the 'ideal modern doctor' Gawande described at the beginning of this book.

They just look to you, patients, to do your part.

Chapter 3: Well Visits
Understanding Screening Tests and Preventive Care

> This chapter addresses well visits. Chapter 2 addressed sick visits.

Millions of Americans get annual physicals that include screening tests. This chapter discusses those tests and suggests some issues to consider and questions to ask your doctor about them.

It does not advise which tests to get or which to avoid.

What are screening tests?

Screening tests look for abnormalities inside of you.

Abnormalities include high cholesterol or blood pressure, growths in a breast or on a prostate and much more. I'll discuss various types of tests and abnormalities in this chapter.

The goal of screening is to identify abnormalities that are small and easy to treat before they grow, become dangerous and are then more difficult and expensive to treat. That's the theory at least.

How well this works in practice is for you and your doctor to decide.

Some quick introductory definitions:

- **Healthy** means a lack of abnormalities. All your indicators are within the norm. You're healthy, in other words, when your doctor can't find anything abnormal about you.
- **Sick or at risk of becoming sick** means the presence of abnormalities. One or more of your indicators are abnormal.
- **Very sick or at high risk of becoming sick** means having abnormalities far from the norm. The farther from the norm, the higher your risk of having a patient event, a heart attack or stroke for example.

Closer to the norm, less risk and less need for medical care; farther from the norm, more risk and more need for medical care.

Screening tests look for abnormalities you can't feel. We call these *asymptomatic* abnormalities. These tests are scheduled on your calendar since, by definition, you can't feel them. You might have an annual mammogram, for example, on May 15, or an annual physical on September 20. Your calendar defines your appointment.

Screening tests differ from diagnostic tests. These look for abnormalities you *can* feel. We call these symptomatic abnormalities.

Diagnostic tests are scheduled according to need. You may call your doctor and complain about chest or stomach pains, for example. Your doctor will tell you to come right in and order diagnostic tests to determine the source of your discomfort, then treat you as determined by the tests.

That discomfort-test-diagnosis-treatment process is a sick visit, discussed in Chapter 2.

This chapter will focus instead on screening tests, i.e. the search for asymptomatic abnormalities.

One cautionary note about screening tests: sometimes they can identify abnormalities that will never cause symptoms or harm. [56] This is called 'overdiagnosis' and comes, in part, from our increasingly powerful screening technologies; we can sometimes identify abnormalities better than we can understand them.

Overdiagnosis is a known consequence of screening tests and can lead to patient harms including unnecessary treatment, diagnosis related anxiety and financial burdens.

Based on this information – that 'sick' or 'at risk' means having abnormalities, and that screening tests sometimes identify non-harmful asymptomatic abnormalities - wise patients discuss three fundamental issues with their doctors.

- **First, the mortality risk question:** Which asymptomatic medical conditions pose high enough risks to justify screening for them?
- **Second, the screening test benefit question:** Which screening tests show enough patient benefit - fewer heart attacks or breast cancer deaths for example - to justify having them?
- **Third, the personal situation question:** At what test result should I (you) do something?

These are extremely difficult questions to answer. I hope this chapter will help you understand them so you can discuss them wisely with your doctors.

The Disease Risk Context

You can't get screened for *every* potential health risk: there are simply too many. Many – most? - are so rare or minor that patients typically ignore them.

Men, for example, don't normally get screened for breast cancer, figuring that the mortality risk is low enough to ignore. But *some* might choose to get screened, suggesting that, in their opinion, the risk is high enough to concern *them.*

Women, on the other hand, more typically do get screened, deeming the mortality risk high enough to concern *them.*

Wise patients, I suggest, determine 'high enough to get screened' for themselves, in conjunction of course, with their doctors. 'High enough' is a personal decision; there's no obvious, objective, universally right or wrong answer.

Where do *you* draw the line? For which disease risks and at what age?

In order to answer that question, you need to estimate your actual risks. One easy-to-use tool to help here is the **Know Your Risk Charts** on the National Cancer Institute website. These estimate your likelihood of dying from various diseases at various ages, over various time periods and, insofar as the government has official disease risk estimates, this would qualify.

I'll present screen shots of various disease risks as examples to show the richness of this information.[57]

First, some disease specific mortality rates for women, then men, aged 30 – 75, showing the number who die per 100 over 10 years. Smokers have *higher* chances of dying from heart attacks, stroke, lung cancer, chronic lung disease and all-causes than shown here; non-smokers, *lower.*

The website listed many more risks but this is a representative sample.

Big Picture Chart for Women, All Races, Ages 30-75 Years
Find the column closest to your age. The numbers tell you the percent of women who will die in the next 10 years from ...

Cause of Death	Age									
	30	35	40	45	50	55	60	65	70	75
All Causes*	0.9%	1.3%	2.0%	3.1%	4.6%	6.5%	9.5%	14.4%	22.5%	35.1%
Vascular Disease										
Coronary Heart Disease	<	0.1%	0.2%	0.3%	0.5%	0.8%	1.3%	2.0%	3.4%	5.9%
Stroke	<	<	0.1%	0.1%	0.2%	0.2%	0.4%	0.7%	1.3%	2.3%
High Blood Pressure[a]	<	<	0.1%	0.1%	0.1%	0.2%	0.2%	0.3%	0.5%	0.9%
Cancer										
Lung and Bronchus	<	<	0.1%	0.2%	0.4%	0.7%	1.0%	1.5%	2.0%	2.2%
Breast	<	0.1%	0.2%	0.2%	0.3%	0.4%	0.5%	0.7%	0.8%	0.9%
Colon and Rectum	<	<	0.1%	0.1%	0.1%	0.2%	0.3%	0.4%	0.5%	0.7%
Ovarian	<	<	<	0.1%	0.1%	0.1%	0.2%	0.3%	0.3%	0.4%
Pancreas	<	<	<	0.1%	0.1%	0.2%	0.2%	0.4%	0.5%	0.6%
Non-Hodgkin Lymphoma	<	<	<	<	<	0.1%	0.1%	0.1%	0.2%	0.3%
Cervical	<	<	<	<	<	<	<	<	<	<
Melanoma	<	<	<	<	<	<	<	<	0.1%	0.1%
Lung Disease										
COPD	<	<	<	0.1%	0.2%	0.4%	0.7%	1.3%	2.0%	2.7%
Asthma	<	<	<	<	<	<	<	<	<	<

Remember when you read this that 0.2% is 2/10ths of 1 percent.

Thus, using 60 year old women as an example, seeing that 2/10ths of 1% die of ovarian cancer over 10 years means 99.8% do not die.

Which presentation has more impact *on you*, learning that 0.2% of 60 year old women die ('I might be the one') or that 99.8% *do not* die?

Consider your own behavioral reaction to each presentation and remember that they're just different ways of saying the same thing.

Now a similar Risk Chart for men. Which risks are high enough to screen for in your opinion?

Big Picture Chart for Men, All Races, Ages 30-75 Years
Find the column closest to your age. The numbers tell you the percent of men who will die in the next 10 years from ...

Cause of Death	Age									
	30	35	40	45	50	55	60	65	70	75
All Causes*	1.7%	2.2%	3.2%	4.9%	7.4%	10.6%	14.6%	20.8%	30.4%	44.6%
Vascular Disease										
Coronary Heart Disease	0.1%	0.2%	0.5%	0.9%	1.4%	2.1%	2.9%	4.2%	6.2%	9.4%
Stroke	<	<	0.1%	0.1%	0.2%	0.3%	0.5%	0.8%	1.4%	2.3%
High Blood Pressure[a]	<	0.1%	0.1%	0.2%	0.3%	0.3%	0.4%	0.5%	0.6%	0.9%
Abdominal Aortic Aneurysm	<	<	<	<	<	<	<	0.1%	0.1%	0.1%
Cancer										
Lung and Bronchus	<	<	0.1%	0.3%	0.6%	0.9%	1.5%	2.2%	2.9%	3.3%
Prostate	<	<	<	<	0.1%	0.1%	0.3%	0.5%	0.8%	1.3%
Colon and Rectum	<	<	0.1%	0.1%	0.2%	0.3%	0.4%	0.5%	0.7%	0.9%
Pancreas	<	<	<	0.1%	0.1%	0.2%	0.4%	0.5%	0.6%	0.7%
Leukemia	<	<	<	<	0.1%	0.1%	0.2%	0.3%	0.4%	0.6%
Esophagus	<	<	<	0.1%	0.1%	0.2%	0.2%	0.3%	0.3%	0.4%
Liver	<	<	<	0.1%	0.2%	0.3%	0.3%	0.3%	0.4%	0.4%
Melanoma	<	<	<	<	<	0.1%	0.1%	0.1%	0.2%	0.2%
Lung Disease										
COPD	<	<	<	0.1%	0.2%	0.4%	0.8%	1.4%	2.3%	3.2%

Finally, my own 10 year mortality risks. I could only fit 17 factors on the screenshot; the website listed over 70.

Your risk of death for
each of the top causes of death from
age **66** until you turn **76** are:

1.	Coronary Heart Disease	4.5%
2.	Lung and Bronchus Cancer	2.4%
3.	COPD	1.7%
4.	Stroke	0.9%
5.	Accidents	0.6%
6.	Colon and Rectum Cancer	0.6%
7.	Prostate Cancer	0.5%
8.	Pancreas Cancer	0.5%
9.	High Blood Pressure	0.4%
10.	Kidney Failure	0.4%
11.	Chronic Liver Disease and Cirrhosis	0.4%
12.	Pneumonia/Flu	0.4%
13.	Septicemia	0.3%
14.	Leukemia	0.3%
15.	Liver Cancer	0.3%
16.	Esophagus Cancer	0.3%
17.	Suicide	0.3%

You can generate a similar chart for yourself. This allows you and your doctor to personalize the discussion and apply this information to your own situation.

There are more charts on the website. It's well worth visiting. I'd offer three take-aways:

First, these are average estimates. You may differ from average in some important way, potentially making these numbers either high or low *for you*. Among potential issues: your specific biology, family history, smoking status, emotional make-up and socioeconomic status. More on the last 2 issues below.

Second, decide for yourself which mortality risks justify screening. Is 0.5% over 10 years high enough? That means 99.5% of people don't die of that disease over 10 years.

Maybe you'll draw the line at 0.8% instead. It's a personal decision. Remember the overdiagnosis risks: drawing the line too low exposes you to more overdiagnosis and treatment risks but drawing the line too high exposes you to more disease risks.

Third, when you learn that a screening test 'cuts your mortality risk by 30%', you can refer back to these charts to determine '30% of what'. This can help you decide if a test works well enough to have. More on that below too.

Understanding the mortality context of various disease risks can enhance conversations with your doctor and help you jointly make wise screening and preventive care decisions.

Let's use this information to revisit that the men vs. women choice of breast cancer screening. According to these charts, 60 year old women face a 0.5% mortality risk over 10 years. Similarly aged white men? Less than 0.1%. We don't know how much less. Here's that Risk Chart.

Cancer Risk Table for Men, All Races, Ages 30-75 Years
The numbers in each age column tell you how many of 1,000 men who will be diagnosed with or die from the cancer in the next 10 years...

Cancer Site	Chance of...	Age									
		30	35	40	45	50	55	60	65	70	75
Breast Cancer	Diagnosis	<	<	<	<	<	<	<	<	1	1
	Death	<	<	<	<	<	<	<	<	<	<

Sources: Risk of developing estimates are based on the SEER 18 areas November 2016 Data Submission.
Risk of dying estimates are based on the US Mortality Files, National Center for Health Statistics, Centers for Disease Control and Prevention.
Risk estimates were calculated for the year range 2012-2014.

< Risk is less than 1 out of 1,000.

Does knowing all this affect your own screening test decisions?

Let's now discuss the two main types of screening tests, lab tests first, and radiology second. As you review this material, try to remember the mortality context. That will help you keep the discussion below in perspective.

Lab Tests
Risk defined by a number

Lab screening tests look for abnormalities in a biological component of your body such as blood pressure, blood sugar or bone density and define abnormalities - or risk - with a number.

If your number is within the normal range, you are healthy.

If your number is outside the normal range, your physician may prescribe a medical intervention to get it back into the normal range.

The numbers that define abnormal change over time, generally to include more people. Initially only 'very abnormal' people were typically defined as at risk, then 'less abnormal', then 'only slightly abnormal' folks, thus increasing the number of people appropriate for medical care.

Each redefinition included people less at risk than previously. Dr. H. Gilbert Welch provided some impact estimates in his 2011 book Overdiagnosed:

Condition	# Sick, Old Definition	# Sick, New Definition	# New Cases
Diabetes (blood sugar decreased from 140 to 126)	11,697,000	13,378,000	1,681,000
High Cholesterol (total cholesterol decreased from 240 to 200)	49,480,000	92,127,000	42,647,000
Osteoporosis (T score decreased from -2.5 to -2.0)	8,010,000	14,791,000	6,781,000
Hypertension (Systolic BP decreased from 160 to 140; Diastolic from 100 to 90)	38,690,000	52,180,000	13,490,000

These expanded disease definitions created over 60 million new patients. How much benefit do the newly defined at risk people get from medical care? A difficult question to answer.

Here's a very rough graph to help us understand how disease risk increases the farther you are from the norm. In this example, total cholesterol levels are the horizontal or X axis; patient events the vertical or Y access. Events include heart attacks, strokes and death.

Disclaimer: The graph below is presented for illustration purposes only and represents guestimates on my part. I do not know the exact shape of this curve. Do not base any medical decisions on it.

Event risk by total cholesterol level per 100 people over 5 years

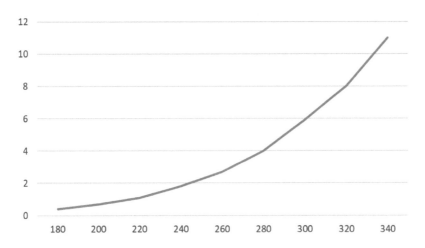

This graph shows that the farther you are from the norm – the higher your total cholesterol in other words - the higher your risk and the more potential benefit from cholesterol lowering medications.

'High cholesterol', in other words, like high blood pressure or high blood sugar or high many other levels, isn't binary. You don't either have it or not. You may have *slightly* high cholesterol, or *fairly* high, or *really* high or somewhere in between. The higher it is, the more at risk you are and the more risk reduction benefit you can enjoy from medical care.

Similarly, the lower your 'slightly high' cholesterol, the lower your risk and the less potential benefit from cholesterol lowering medications.

Doctors often want to define 'high' cholesterol or high other levels expansively to include as many patients as can possibly benefit in the definition. Over time the definition of high total cholesterol fell from 240 to 200 as studies showed that people at lower levels of 'high cholesterol' faced event risks and could benefit from cholesterol lowering medications.

Pharmaceutical companies similarly want to expand disease definitions so they can increase the markets for their medications. There may be too few far-from-the-norm, very-at-risk patients available to satisfy drug manufacturer business needs.

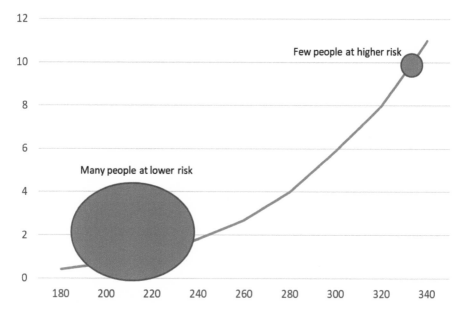

Largely for these two reasons – doctors wanting to provide as much benefit as possible and drug companies and similar product providers wanting the largest possible market – at risk definitions have expanded over time.

Remember that lab tests are indicators. They only suggest something about your likelihood of having a medical event. That correlation, between indicators and events, may be strong or weak, depending on the specific test.

I'd encourage people to review Chapter 2 at this point, especially the 'out of 100 people like me, how many benefit and are harmed by this treatment?' question and the related discussions of indicator vs. patient event benefits and medical guidelines. Both topics are worthy of discussion with your doctor if/when you are diagnosed with 'high' something.

Various organizations develop guidelines for treating high or low of some indicator to help doctors and patients understand the numbers and make wise treatment decisions. Though some guidelines may be very helpful, others can confuse patients and physicians.

Here's a brief case study of hypertension (high blood pressure) guideline confusion to highlight the need to discuss all this with your doctor and not to follow any guideline blindly.

First the American Heart Association's guidelines published in their journal Hypertension on November 13, 2017.

KNOW YOUR BLOOD PRESSURE
—AND WHAT TO DO ABOUT IT

By AMERICAN HEART ASSOCIATION NEWS

Systolic

Diastolic

<120 mmHg
—— AND ——
<80 mmHg

120-129 mmHg
—— AND ——
<80 mmHg

130-139 mmHg
—— OR ——
80-89 mmHg

≥140 mmHg
—— OR ——
≥90 mmHg

The newest guidelines for hypertension:

NORMAL BLOOD PRESSURE
*Recommendations: Healthy lifestyle choices and yearly checks.

ELEVATED BLOOD PRESSURE
*Recommendations: Healthy lifestyle changes, reassessed in 3-6 months.

HIGH BLOOD PRESSURE / STAGE 1
*Recommendations: 10-year heart disease and stroke risk assessment. If less than 10% risk, lifestyle changes, reassessed in 3-6 months. If higher, lifestyle changes and medication with monthly follow-ups until BP controlled.

HIGH BLOOD PRESSURE / STAGE 2
*Recommendations: Lifestyle changes and 2 different classes of medicine, with monthly follow-ups until BP is controlled.

*Individual recommendations need to come from your doctor.
Source: American Heart Association's journal Hypertension
Published Nov. 13, 2017

But second, the US Preventive Services Task Force – remember them from Chapter 1 – recommends treating people **over age 60 to a target blood pressure of 150/90 mm Hg** and **under age 60 to a target 140/90** though 'some experts believe that this should also be maintained in those aged 60 years or older' according to their write-up. [58]

Their targets, either 140 or 150/90, are the American Heart Association's 'very high'.

At the same BP numbers, the USPSTF says 'good' and the AHA says 'take 2 different classes of medicine'.

Who's right? At what level should *you* take action? Difficult questions.

Clearly 240/190 is dangerous blood pressure. That's the great benefit of this screening test, identifying people at serious risk from this oft-called silent killer. But there aren't many people at that high level; many more are in the 125 – 160 systolic range. (Systolic blood pressure is the top number.) That's where most of the 'when should I take it?' discussion resides.

My suggestion here: you should take medications at the number that you, a wise, well informed patient, and your doctor, a caring expert, agree on. I'll present some issues to consider below, along with a case study of my own blood pressure situation to articulate this issue more fully.

Then, after you decide *when* to take medications, *which* should you take? There's disagreement about that too.

I encourage people to review Chapter 1, especially the discussion of Cochrane to address that question. Cochrane posts systematic reviews like these:

- Hydralazine for treatment of high blood pressure
- Eplerenone for high blood pressure

- Methyldopa reduces blood pressure in people with high blood pressure
- Spironolactone for the treatment of high blood pressure
- Beta-blockers for hypertension
- Thiazides best first choice for hypertension
- And more.

A few minutes reviewing these meta-studies helps you determine which treatments work well, which poorly and which in between. Knowing that can help you and your doctor have a more productive interpretive discussion.

One additional comment about current guidelines for treating high blood pressure, cholesterol, blood sugar and the like. Some seem to favor medical interventions over lifestyle changes, at least as measured by the number of words or pages allocated to each.

The 2018 American College of Cardiology / American Heart Association 'Guideline on the Management of Blood Cholesterol', for example, allocated one paragraph out of 120 pages to diet. [59]

By contrast, the US National Institutes of Health publish very detailed dietary and lifestyle advice such as the 2005 'Your Guide to Lowering Your Cholesterol with TLC – Therapeutic Lifestyle Changes'.[60]

How might your doctor's choice of guidelines affect his or her treatment recommendations?

I encourage patients to remember Atul Gawande's 'interpretive' ideas from the beginning of this book and to treat guidelines as suggestions while they discuss a wide range of treatment options with their physicians, not just medical interventions.

Let's summarize what we've learned so far before moving on.

- The farther your indicators are from the norm, the higher your chance of having a patient event.
- Consequently, the farther you are from the norm, the more medicine can reduce your risk.
- Physicians and private sector suppliers want everyone who could possibly benefit from medical care to have access to it,

thus expanding the definition of 'at risk' over time and potentially exposing patients to more overdiagnosis risks.
- The expanded definitions increasingly include lower risk people.
- Different organizations promulgate different guidelines for dealing with lab screening test results.
- Then different studies suggest different ways to treat the discovered abnormalities.

Tread thoughtfully through all this with your doctor.

****

Patients can have lots of different abnormalities (a.k.a. risk factors), not only cholesterol and blood pressure, but also blood sugar, body-mass-index, blood oxidation rate, lung capacity etc. You might be normal on some, slightly abnormal on others and really abnormal on still others.

How does a wise doctor proceed with different patients? Consider the simple comparison below. Should the doctor treat both the same?

Patient A	**Patient B**
BP 145 / 85	BP 145 / 85
Non smoker	Smoker
1 family member had coronary disease	No family history of coronary disease
Normal cholesterol, slightly high blood sugar	Normal cholesterol and blood sugar
Appropriate Body Mass Index	Slightly high Body Mass Index

To answer these questions in the coronary arena, the American College of Cardiology and American Heart Association together developed a Risk Calculator in 2013 that includes:

- Age
- Gender
- Smoker
- Diabetes

- Systolic blood pressure
- Total cholesterol
- HDL
- Chronic kidney disease
- Family history
- Currently on blood pressure lowering medications or not

Physicians enter values for all these factors into an algorithm that predicts the patient's 10-year likelihood of having a coronary event.

Now physicians can answer the question above, whether or not to treat Patients A and B similarly. The answer depends on each person's 10-year event likelihood.

These two organizations then go one step further. They recommend patients begin statin therapy when their 10-year coronary event likelihood exceeds 7.5% and they're between ages 40 – 75. [61]

If Patient A has a 6% 10-year likelihood, don't treat. But if Patient B has a 12% likelihood, treat.

Bingo. Problem solved.

Almost.

Researchers have found three major problems with the Risk Calculator. These problems plague the updated 2018 version too.

First, this calculator, as all risk calculators, has a margin of error of 5% *at best*. [62] 'Margin of error' means the calculated risk isn't exact but rather an estimate. 'At best' means the actual margin might be more than 5%.

'At best' also means we don't know the real margin of error!

Let's put some hypothetical numbers to this. Let's say your calculated 10-year risk of having a coronary event is 11%. The '5% margin of error' means your real 10-year risk is somewhere between 6% and 16%, i.e. somewhere between 'treat with statins' and 'don't treat with statins.'

But what if the margin of error is 10%? It might be - we don't know. Then your calculated 10% risk means your actual risk is somewhere between about 1% and 20%. Doctors would clearly make very different treatment recommendations at those two extremes, and patients very different decisions.

You want to avoid being *over*treated, which exposes you unnecessarily to treatment harms, and *under*treated which exposes you unnecessarily to disease harms. Clearly something to discuss with your doctor.

Second, researchers have tested this calculator against real life patients. In one study of 65+ year olds with no coronary disease but moderate hyperlipidemia and hypertension, for example, no benefit was found when pravastatin was given for primary prevention. [63]

But the calculator said 'treat'.

Third, the Risk Calculator leaves out any mind-body and socio-economic factors, all of which, research suggests, impact disease rates.

Some research on loneliness, depression and stress, for example: [64]

- Loneliness is a risk factor for coronary artery disease according to the Oxford Research Encyclopedias, 2017

- People 60 years and older who reported struggling with loneliness faced an increased risk of mortality compared with participants who do not report being lonely according to a study in the Archives of Internal Medicine, July 2012

- Psychosocial factors like depression and stress are as strong risk factors as high blood pressure and nearly as important as diabetes according to the 2004 INTERHEART study of 25,000 patients in 52 countries. [65]

Former US Surgeon General Vivek Murthy put it this way in a Harvard Business Review article:

> During my years caring for patients, the most common pathology I saw was not heart disease or diabetes; it was loneliness. [66]

But loneliness, depression and stress are not included in the Risk Calculator or typical medical guidelines.

Nor is socioeconomic status. Here's Angus Deaton, Nobel Prize winning economist who has studied disease patterns summarizing his research:

> The finding that income predicts mortality has a long history... The mortality gradient by income is found wherever and whenever it is sought.

The higher your income, Deaton and others have found, the lower your disease and death rates.

That's perhaps not very surprising: wealthy people eat healthier food, smoke less and visit doctors more.

But the really impressive finding is that the same person – same cholesterol, smoking status, body mass index etc. – faces different disease and mortality risks depending on income.[67] "There's something about lower socioeconomic status itself that increases the risk of premature death" and of various diseases according to the New England Journal of Medicine's 2004 analysis.[68]

And the reverse. The higher you are on the socioeconomic ladder, the lower your disease and death risks. Some studies suggest that low socioeconomic people are 3x more likely to have medical problems and die of heart disease than high income folks.[69]

"An individual's health can't be torn from context and history," says Harvard Professor Madeline Drexler. "We are both social and biological beings...and the social is every bit as real as the biological."[70]

But none of these factors – loneliness, socioeconomic status or similar – enter the College of Cardiology and Heart Association Risk Calculators. Try to remember that when you and your doctor next look at your numbers.

The question is less 'What do the guidelines say?' than 'At what number should I, with my own medical makeup, family history, loneliness situation, economic status and stress factors do something?' That's a very different question and one far more difficult to answer.

I hope the discussion so far will help you and your doctor together determine the answer.

At what blood pressure level should I, Gary Fradin, take medications?

I'll offer a brief, personal case study as an academic example to tie many of these issues together.

I present this as an exercise for patients to undertake in advance of lab tests. This can help take the emotion out when you learn that you're high or low on some scale.

I'm 66 years old, happily married, financially comfortable though far from rich, generally happy and pretty satisfied with life.

My Body Mass Index is around 29 and total cholesterol 210 when last checked a couple years ago. I don't smoke and probably don't have much family history of coronary disease; though both my father and maternal grandmother developed it, they both developed it well after age 60.

I exercise regularly, about 4 hours per week of cardiovascular and resistance training in my local YMCA.

And, mainly because of things I study for a living, I worry more about being *over*treated, i.e. side effects of treating harmless abnormalities, than being *under*treated, harms from an untreated asymptomatic abnormality. That's my own personal preference; it may not be yours.

Based on all this, what is high blood pressure *for me*? At what point should I take medications to lower it?

(Pause here while you come up with the number.)

(Still pausing so the discussion below doesn't influence you.)

I think my disease risk is lower-than-average so, in my opinion, I should start blood pressure medication at a higher-than-typically-recommended level, probably around 160/100 or so.

I'd make a very different decision if I was depressed, lonely, highly stressed, financially insecure or lacked the energy to exercise. In that case, I might start medication at 130/80 or even slightly lower.

Do you agree? Disagree? There's no one objectively right answer here.

I would, of course, discuss all this with my doctor and listen carefully to his input. That's why it's so important to choose your doctors wisely: you want someone who can engage with you on a wide range of issues including emotion, social connectedness, risk orientation, fears and hopes, not simply numbers.

Indeed, some doctors think the clinical interview, not the actual lab test results, is the most important part of the check up.

Do you agree?

Does your doctor?

Radiological Screening Tests

Now let's turn to the second kind of screening test, radiological scans. These look for asymptomatic abnormalities with images, things like cancerous growths.

I'll introduce this discussion with a 'play radiologist' exercise.

Instead of looking at radiological scans of the human body, though, that are really difficult to interpret – one author called his book on this subject Snowball in a Blizzard: A Physician's Notes on Uncertainty in Medicine [71] - we'll look at maps. And instead of looking at cancer tumors, we'll look at lakes. In other words, a lake for our purposes is a cancerous tumor. [72]

Wikipedia defines lake as an area filled with water, localized in a basin, that is surrounded by land. I just want to clarify that before we begin to avoid confusion later on.

How many lakes do you see on this map? For simplicity purposes, just look at Florida.

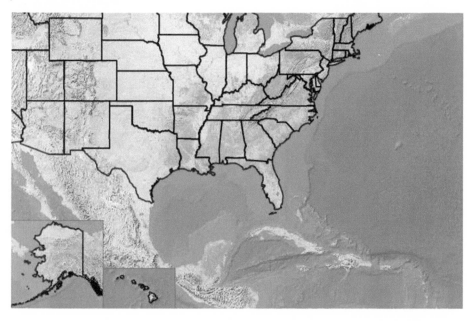

Most people say 'one'. Fair enough. You can't cheat here and say 'I want to see the map more clearly.' This map resembles what a radiologist might actually see, or might have seen with old technology.

Let's now upgrade our technology to look more closely. How many lakes in this map of central Florida?

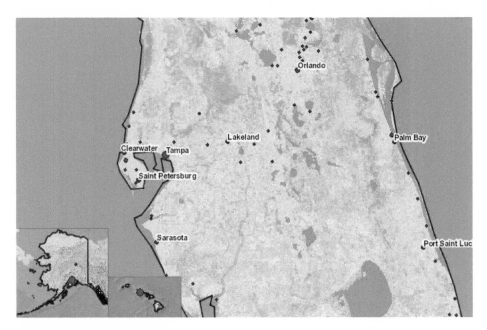

Most people say 'many'.

We haven't changed Florida. We've just used better technology. The actual number of lakes remains the same.

Even better technology finds this.

It fits Wikipedia's definition: an area filled with water, localized in a basin, that is surrounded by land so it must be a lake. But most people would disagree, saying it's not really a lake, it's a swimming pool which has some important different characteristics.

In other words, as our technology improves, we find more abnormalities and need to alter or re-interpret their definitions.

That, in a nutshell, is the problem radiologists encounter when they look for asymptomatic abnormalities. They find things that fit the definition but that may not be harmful. This leads to the 'overdiagnosis' problem we discussed earlier.

Let me give a non-cancer example, MRI scans of professional baseball pitcher shoulders. [73] Dr. James Andrews, a well known sports medicine orthopedist, scanned shoulders of 31 healthy pitchers. These guys were not injured and had no shoulder pain complaints.

But the MRIs found abnormal rotator cuff tends in 27 of them!

Swimming pools, apparently.

But ask yourself, what would go through a pitcher's mind once he learns these MRI results? Could it affect his psyche and lead to bad on-field performances? Could it lead to (unnecessary) rotator cuff surgery and an entire missed season? Plenty of potential overdiagnosis downsides. Any compensating upsides?

We see a similar overdiagnosis situation in cancer screening. In fact, overdiagnosis of harmless abnormalities is one reason for the current epidemic of various cancers: radiologists find more swimming pools these days than previously because of our powerful technologies. We can identify abnormalities sometimes better than our scientific knowledge allows us to differentiate dangerous from harmless ones.

Dr. Jill Wruble, radiologist and Assistant Professor at Yale offers a useful insight here, suggesting that patient survival depends more on the tumor's biology rather than the point at which it is discovered.[74] Unfortunately we're not terribly good at tumor-biology-identification yet so we fall back on the less strongly predictive tool, early detection and find ourselves in the 'swimming pool or aggressive cancer' conundrum.

The most extreme example of this comes from South Korea which instituted widespread thyroid screening in the late 1990s. This chart shows the resulting thyroid cancer incidence – the amount of cancer identified – and mortality, the rate at which people die of thyroid cancer.

You can see that incidence skyrocketed – that's the bright line that rises sharply after about 2000 - while mortality remained about the same, the bold horizontal line at the bottom. [75]

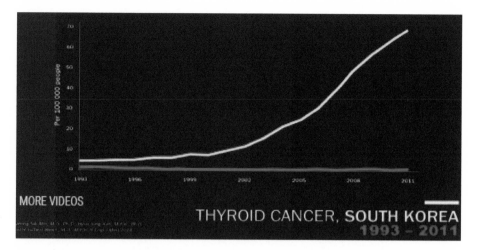

Once identified, patients generally want treatment. That can raise their risks of treatment harms unnecessarily; if the abnormality wasn't dangerous to begin with, then the care wasn't necessary.

Gilbert Welch puts this more succinctly: 'an overdiagnosed patient can't benefit from treatment' because the patient wasn't sick in the first place.

Overdiagnosis from excessive screening with new, really powerful technologies has several implications, most pretty obvious – unnecessary care, unnecessary anxiety, unnecessary medical expenditures – but I want to suggest a surprising one: higher cancer incidence in more affluent communities. How can that be? More affluent people are healthier according to Angus Deaton earlier in this chapter. They should have less cancer, not more.

As this chart from Welch's 2017 study in the New England Journal of Medicine shows, incidence of 4 different cancers – breast, prostate, thyroid and melanoma – is higher in wealthier counties.[76] But disease specific mortality rates are not lower.

Welch defined wealthy counties as those with median family income above $75,000 and poor as median incomes below $40,000. This particular study looked at caucasians only to avoid racial or genetic bias and calculated disease specific mortality rates, not overall mortality.

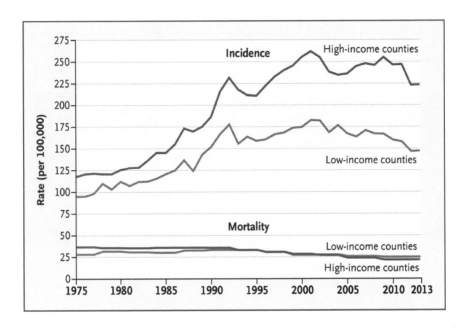

All this raises the question of patient benefit from easier access to cancer screening tests, and consequently more tests. Can you see why? More testing can lead to more overdiagnosis leading in turn to more risk of side effect harms, infections, errors etc. without any mortality reduction.

Let's summarize:

- Radiological tests look for abnormalities, often growths.
- These are very difficult to identify precisely.
- The more powerful our radiological technologies, the smaller the abnormalities we can identify.
- Abnormality definitions expand as we identify smaller and smaller abnormalities.
- The more we screen, the more abnormalities we find, but not necessarily the lower the disease specific mortality rates.

Again, as with lab tests, tread thoughtfully here.

Who can help me decide which screening tests to have and what to discuss with my doctor?

As suggested in Chapter 1, two reliable and thoughtful information sources are the US Preventive Services Task Force and Cochrane.

I'll present USPSTF and Cochrane recommendations and analyses below for breast and prostate cancers, and the USPSTF recommendation for thyroid cancer as examples. Note the similarity in their conclusions.

As you read through them, consider two key issues:

- How do their recommendations complement or diverge from the Risk Chart data introduced at the beginning of this chapter?
- How do these recommendations affect your own medical decision making and discussions with your doctors?

First, a screen shot of the USPSTF breast cancer /mammography recommendations. I cut, pasted and deleted part of this for space reasons here. You can find their entire recommendation statement at

https://www.uspreventiveservicestaskforce.org/Page/Document/Rec ommendationStatementFinal/breast-cancer-screening1

Final Recommendation Statement
Breast Cancer: Screening

Recommendation Summary

Population	Recommendation	Grade (What's This?)
Women aged 50 to 74 years	The USPSTF recommends biennial screening mammography for women aged 50 to 74 years.	B
Women aged 40 to 49 years	The decision to start screening mammography in women prior to age 50 years should be an individual one. Women who place a higher value on the potential benefit than the potential harms may choose to begin biennial screening between the ages of 40 and 49 years.	C
Women aged 75 years or older	The USPSTF concludes that the current evidence is insufficient to assess the balance of benefits and harms of screening mammography in women aged 75 years or older.	I
All women	The USPSTF concludes that the current evidence is insufficient to assess the benefits and harms of digital breast tomosynthesis (DBT) as a primary screening method for breast cancer.	I
Women with dense breasts	The USPSTF concludes that the current evidence is insufficient to assess the balance of benefits and harms of adjunctive screening for breast cancer using breast ultrasonography, magnetic resonance imaging, DBT, or other methods in women identified to have dense breasts on an otherwise negative screening mammogram.	I

These recommendations apply to asymptomatic women aged 40 years or older who do not have preexisting breast cancer or a previously diagnosed high-risk breast lesion and who are not at high risk for breast cancer because of a known underlying genetic mutation (such as a BRCA1 or BRCA2 gene mutation or other familial breast cancer syndrome) or a history of chest radiation at a young age.

The USPSTF gives screening mammography a B for women 50 - 74 and recommends biennial screening not annual.

They justify the B recommendation quantitatively, estimating that screening 10,000 women in each age group over ten years results in:

- 21 fewer breast cancer deaths for women 60 – 69
- 8 fewer breast cancer deaths for women 50 – 59
- 3 fewer breast cancer deaths for women 40 – 49

The Task Force also discussed the potential harms, mainly overdiagnosis, and estimates that between one-in-three and one-in-five women diagnosed with breast cancer through annual mammography gets treated for cancer that would never have been discovered or caused health problems in the absence of screening.

Biennial screening brings this down to 1/8 women with only a small incremental decrease in the number of breast cancer deaths averted. Even at the 1/8 rate, they estimate that for every breast cancer death avoided, between 2 and 3 women will be treated unnecessarily. (Remember the unnecessary treatments have their own side effect risks.)

Let's compare the USPSTF recommendations to Cochrane's 2013 write up Screening for Breast Cancer with Mammography. Here's a screen shot of Cochrane's introduction. The underlined words are hot links on their website.

Screening for breast cancer with mammography

Screening with mammography uses X-ray imaging to find breast cancer before a lump can be felt. The goal is to treat cancer earlier, when a cure is more likely. The review includes seven trials that involved 600,000 women in the age range 39 to 74 years who were randomly assigned to receive screening mammograms or not. The studies which provided the most reliable information showed that screening did not reduce breast cancer mortality. Studies that were potentially more biased (less carefully done) found that screening reduced breast cancer mortality. However, screening will result in some women getting a cancer diagnosis even though their cancer would not have led to death or sickness.

The higher quality studies found no breast cancer mortality benefit from mammography but the lower quality studies did find a benefit. As with all these ambiguous conclusions, the ultimate decision is yours. That's why having a good interpretive physician is so important.

Let's turn to prostate cancer screening recommendations. Here's the USPSTF recommendation for prostate cancer screening, again cut, edited and pasted for space purposes here. You can read the entire write up here

https://www.uspreventiveservicestaskforce.org/Page/Document/Rec ommendationStatementFinal/prostate-cancer-screening1

Prostate Cancer: Screening

Release Date: May 2018

Population	Recommendation	Grade (What's This?)
Men aged 55 to 69 years	For men aged 55 to 69 years, the decision to undergo periodic prostate-specific antigen (PSA)–based screening for prostate cancer should be an individual one. Before deciding whether to be screened, men should have an opportunity to discuss the potential benefits and harms of screening with their clinician and to incorporate their values and preferences in the decision. Screening offers a small potential benefit of reducing the chance of death from prostate cancer in some men. However, many men will experience potential harms of screening,	C
Men 70 years and older	The USPSTF recommends against PSA-based screening for prostate cancer in men 70 years and older.	D

Cochrane says pretty much the same thing in its 2013 review Screening for Prostate Cancer. Here's the author conclusion:

Authors' conclusions:

Prostate cancer screening did not significantly decrease prostate cancer-specific mortality in a combined meta-analysis of five RCTs. Only one study (ERSPC) reported a 21% significant reduction of prostate cancer-specific mortality in a pre-specified subgroup of men aged 55 to 69 years. Pooled data currently demonstrates no significant reduction in prostate cancer-specific and overall mortality. Harms associated with PSA-based screening and subsequent diagnostic evaluations are frequent, and moderate in severity. Overdiagnosis and overtreatment are common and are associated with treatment-related harms. Men should be informed of this and the demonstrated adverse effects when they are deciding whether or not to undertake screening for prostate cancer. Any reduction in prostate cancer-specific mortality may take up to 10 years to accrue; therefore, men who have a life expectancy less than 10 to 15 years should be informed that screening for prostate cancer is unlikely to be beneficial. No studies examined the independent role of screening by DRE.

Finally, the USPSFT on thyroid cancer screening since I showed the South Korea experience above.

Thyroid Cancer: Screening

Release Date: May 2017

Recommendation Summary

Population	Recommendation	Grade (What's This?)
Adults	The USPSTF recommends against screening for thyroid cancer in asymptomatic adults.	D

The Task Force determined with moderate certainty that screening for thyroid cancer in asymptomatic persons results in harms that outweigh the benefits. You can read their entire statement here

https://www.uspreventiveservicestaskforce.org/Page/Document/Rec ommendationStatementFinal/thyroid-cancer-screening1

A closing thought

You can search the USPSTF and Cochrane websites for discussions about other types of screening tests. Again, as suggested in Chapter 1, these sites provide high quality, objective analyses and generally come to similar conclusions. That increases my confidence in them.

Are these recommendations and conclusions compelling? Encouraging? Discouraging? That's for you and your doctor to interpret as appropriate for you.

How closely do they mirror the Risk Chart data presented earlier in this chapter? Again, something for you and your doctor to discuss.

Let me close by posing the fundamental screening question. I introduced this above in the discussion about my own blood pressure.

It's pretty complicated so read it slowly.

Are you more concerned about having an asymptomatic abnormality that becomes lethal before it becomes symptomatic (in other words, a medical problem that you can't feel until it is too late to treat)....**or about the side effects of treating a harmless abnormality?**

It's a tough question to answer because you'll have to weigh the benefits and risks of screening against the benefits and risks of not-screening, probably on a test-by-test basis.

It also underscores the importance of doing your own homework (see Chapter 1), of choosing your doctor(s) wisely and of having a good working relationship with him/her or them.

Good luck!

Gary Fradin

Chapter 4: How to read a medical news article

Chapter 1 of this book introduced three reliable patient information sources: Cochrane, ChoosingWisely and the US Preventive Services Task Force. **As a general and practical rule-of-thumb, I suggest staying with them and skipping this chapter.**

But some patients may want to learn more about a specific medical condition or therapeutic options. Others may see interesting medical news headlines online or in newspapers.

Yet others don't believe my rule-of-thumb recommendation above!

This chapter presents some tools to use when doing your own research.

A warning before we start: this discussion is pretty complicated. Or, stated differently, reading a medical news article critically is way harder than most people think.

Let's start with a case study example to demonstrate, a story on CNN.com entitled **'Not exercising worse for your health than smoking, diabetes and heart disease, study says'** by Wayne Drash, published on October 20, 2018.[77] This is not a particularly interesting or egregious article but simply one that I recently read.

I'll use it to introduce some key 'wise reader' issues: correlation vs. causality, inappropriate vs. appropriate ways to report medical risks and the divergence between eye-catching headlines and article substance.

I'll show how an untrained reader can be misled by uncritically presented information about a low-quality study.

Here's the gist of Drash's CNN article:

> A new study found that a sedentary lifestyle is worse for your health than smoking, diabetes and heart disease.
>
> One of the study's authors, Cleveland Clinic cardiologist Dr. Wael Jaber said "Being unfit on a treadmill or in an exercise

stress test has a worse prognosis, as far as death, than being hypertensive, being diabetic or being a current smoker."

Researchers retrospectively studied 122,007 patients who underwent exercise treadmill testing at the Cleveland Clinic between January 1, 1991 and December 31, 2014 to measure all-cause mortality.

The risk associated with death is "500% higher" for those with a sedentary lifestyle than for the top exercise performers according to Dr. Jaber.

The actual CNN article is longer but makes essentially the same points: the better you score on exercise stress tests, the lower your chance of dying. This all sounds pretty reasonable and looks straightforward to most people.

But it isn't.

Drash's article and Jaber's quotes make the underlying study sound like a comparative study of two groups of people, one of which exercised and the other of which did not.

Or maybe three groups: one that exercised and smoked, one that exercised and didn't smoke and one that didn't exercise or smoke.

But it wasn't a comparative study; it didn't compare two or three groups of people. Instead it was an *observational* study that looked only at one group of people.

Observational studies look for correlations and attempt to draw conclusions. These are always weak and unreliable because correlation does not imply causality.

- Correlation means 'some people had factor A and event B.'
- Causality means 'factor A caused event B.'

We determine causality from comparative studies as discussed in the previous chapters. That's also called the scientific method, practiced since Galileo in physics, chemistry, biology and sometimes medicine.

We determine correlation in lots of different ways and sometimes discover correlations that don't make sense. Take, for example, an observational study of shoe size (factor A) and reading comprehension scores (event B). Shoe size correlates pretty closely with reading comprehension.

Yes, it actually does.

Infants and toddlers who have very small feet don't read as well as older people who have bigger feet.

Shoe size obviously didn't cause poor reading comprehension and a headline suggesting it does is nonsense: 'Small feet worse for reading comprehension than bad teachers.'

Compare that to Drash's article title 'Not exercising worse for your health than smoking...' based on similarly observed correlations.

I'll point out three problems in Drash's article, two based on the observational study methodology introduced above and one based solely on presentation, all of which put the conclusion in question and none of which were addressed in the article itself:

- First, this study might have assumed or suggested causality backward. Some people, let's call them 'healthier', can exercise; their bodies allow it. Other people, we'll call them 'unhealthier', might not be able to exercise; their physiological make-up precludes it. They might have inherently poor lung capacity, weak hearts or poor blood oxidation rates ... who knows the reason.

  These 'unhealthier' people face higher mortality rates as a result of their underlying biology. They don't exercise because their bodies don't allow it.

  This study, therefore, might only have shown that 'healthier' people i.e. those able to exercise, do, in fact, exercise more than 'unhealthier' folks i.e. those unable to exercise.

  The underlying biology, not exercise, affects mortality rates. 'Unhealthier' people die younger. We already knew this.

- Second, remember our discussion in Chapter 3 about loneliness and socio-economic status impacts on disease and death rates. Lonely people face higher mortality rates, as do low socio-economic folks. We know this from other studies.

  Might lonely people, or socioeconomically disadvantaged ones, exercise less than friendlier, higher income folks? Maybe.

  Might this study simply have discovered, or confirmed, that lonely or low-income people exercise less? We don't know because it was an observational, not comparative study.

  And, if you really want to push this, might 'unhealthy' (from our discussion above), lonely, socio-economically disadvantaged people exercise less than healthy, friendly, middle class folks? Maybe. Perhaps that's the sedentary lifestyle group.

  This study might therefore simply have confirmed some sociological patterns, not broken any medical news.

- Third, this article followed a standard format of including some sensational sounding quotes. Those may sell website clicks and ads more than inform patients. See, for example Dr. Jaber's comment about the risk of death in people associated with a sedentary lifestyle being 500% higher than for the fittest people. That sounds startling and impressive.

  But...500% higher *than what*? Remember the discussion in Chapter 2 of inappropriate answers to 'out of 100 people like me, how many benefit and are harmed?' Here the wise reader asks '500% higher *than what*?' The article doesn't tell us.

A more well-written article would have pointed out these flaws and a really well-written article would have addressed them.

Wise patients who read this article critically end up wondering how bad not-exercising really is. They may well conclude that this article doesn't tell us and that there's less here than meets the eye.

By-the-way, exercise probably is good for you. I got that information from other sources, not this article.

****

Let's put these issues into context: Gary Schwitzer and his team at HealthNewsReview (www.HealthNewsReview.org that unfortunately ceased operations in December, 2018) evaluated over 2600 medical news articles over 12 years in news outlets like the New York Times, NPR, the Wall Street Journal, Bloomberg News, NBC, CBS, Fox, ABC, CNN, Time, Newsweek and similar.

They found that

- 66% of stories did not quantify benefits
- 63% did not quantify harms
- 62% did not evaluate the evidence quality.

Schwitzer also found that many medical news articles emphasize or exaggerate benefits with words like 'important', 'significant', 'breakthrough' and 'game changer' but ignore or minimize harms with words like 'infrequent' and 'minor'. Breakthrough and minor mean different things to different people so tend to convey emotions rather than facts. These may or may not serve your own educational purpose.

(See why I like Cochrane and the US Preventive Services Task Force so much?)

I'll introduce below five key questions that a good medical news article addresses to help readers avoid common pitfalls. I'll also provide examples of both good and bad articles.

- Out of 100 people, how many benefit and how?
- Out of 100 people, how many are harmed and how?
- Does the article discuss the evidence quality?
- Did the article explain any conflicts of interest?
- Does the article relate new information to existing?

Out of 100 people, how many benefit and how?

Wise readers always look for the number of people who benefitted and an indication of the types of benefit measured.

Wise readers similarly are skeptical of words like 'game changer', 'breakthrough', 'promising', 'dramatic' or similar. These tend to convey emotions rather than hard data.

- Does a game changing therapy benefit 3 more per 100 or 37 more than the old therapy? There's no standard definition.
- Does a promising drug improve patient outcomes or simply affect an indicator? Remember our discussion in Chapter 2 about Niaspin. That's just one example where the indicator benefit doesn't alter patient event rates.
- Does a dramatic benefit mean a very sick patient could suddenly walk again or simply drink a cup of water? Again, no standard definition.

Here's a medical news article that addresses the benefit issue well, from the New York Times '**Taking Fish Oil During Pregnancy is Found to Lower Child's Asthma Risk.**' [78] I have no particular interest in fish oil, only an interest in showing how a good article describes benefits.

Among children whose mothers took fish-oil capsules, 16.9 percent had asthma by age 3, compared with 23.7 percent whose mothers were given placebos.

(Remember that 16.9% means 16.9 per 100.)

Let's put that sentence into our comparative study graphic.

**Control Group
100 people

23.7 had asthma
by age 3**

**Treatment Group
100 people

16.9 had asthma
by age 3**

We easily see the benefit of fish oil capsules: 6.8 children per 100 avoided asthma. I don't know if this is a game changer or dramatic, but it may be worth discussing with your doctor if you're a pregnant woman.

Here's a second example of a good article, this time that quantifies the benefit improvement, **'In a small Alzheimer's study, hints of modest benefit'** from STAT. [79]

> the patients who received all six low doses did the best; their average improvement after the 12 weeks was 1.5 points on the 100-point cognitive test.
>
> Patients receiving a placebo ... lost 1.1 points.
>
> The results were not statistically significant.

Here we didn't learn *how many* patients improved but instead *how much* they improved. Again, our comparative study graphic:

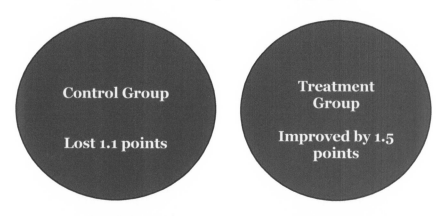

We can conclude that patients who took this medication benefited by 2.6 points, the difference between the control and treatment groups...or can we?

This article said the results were not statistically significant. Time to define 'significant' in the two commonly used but very different ways.

- **Statistically** significant means the result didn't happen by accident or wasn't a fluke. In common use, statistical significance means that another research study, using the same methodology, would likely generate similar results.

  'Not statistically significant' means the results may have happened by chance and another research study, using the same methodology, might not generate similar results.

  Statistical significance applies to the study methodology and data quality.

- **Clinically** significant means that it impacted a large enough number of patients to warrant physician attention. An outcome that benefits 1 patient in 10,000 probably isn't clinically significant but one that benefits 19 in 100 probably is.

Thus, a statistically significant study might not be clinically significant.

In the case of this Alzheimer's drug, the results were not statistically significant, meaning the researchers are not very confident that another set of researchers looking at the same drug and patient population would generate similar conclusions.

It may or may not be clinically significant. That's for you and your doctor to decide together. Some physicians may say the impact is too small to be a game changer. (!)

Now for an example of an article that defines benefits poorly, **'PSA screening for prostate cancer saves lives after all, study says'** from the Los Angeles Times. [80] Screening men over age 55, this article says

> can significantly reduce the risk for prostate cancer death. (Which definition of 'significant', statistical or clinical?)

When men who fit the criteria for screening get the PSA test, the reduction in deaths due to prostate cancer was between 25% and 32%, the new study found.

Between 25% and 32% *of what*? We don't know and can't put this information into our comparative study graphic format, the two circles I used above. We don't know how many benefited per 100 men.

The article is misleading because a 25% mortality reduction sounds very impressive but may not be.

I looked up *of what* in this case using the Risk Charts from Chapter 3. A 55-year old man has a 0.1 in 100 chance of dying of prostate cancer over the next 10 years.

A 25% mortality reduction from 0.1 in 100 of men over 10 years reduces the mortality risk by about 0.025 men per hundred. That's a quarter of a tenth of a man.

Note the emotional impact of this article's risk reduction statement on readers:

- A 25 – 32% mortality reduction sounds very impressive.
- A quarter of a tenth of a man per 100 mortality reduction sounds less impressive.
- But they're the same number!

That's why this was a poor article. It confused more than it illuminated and required the reader to do so much additional research in order to determine the actual screening test benefit. (The article did appropriately note some screening test harms. More on harms below.)

I hope you're beginning to see why you need to read medical news articles so critically.

Out of 100 people, how many are harmed and how?

Risks always exist in medical interventions. A good article explains the risks; a poor one doesn't.

Here's an example of a good article, **There's Something in Magic Mushrooms That's Shown to Ease Anxiety and Depression in Cancer Patients in One Dose**, again by Healy in the Los Angeles Times.[81] Parenthetically this shows how the same reporter can write some good and some poorer articles.

> In the two trials, about 15% of subjects experienced nausea or vomiting when getting a high dose, and about 1 in 3 experienced some form of transient psychological discomfort.

> Many subjects' heart rates and blood pressure rose, but none to a dangerous extent.

Back to our comparative study graphic.

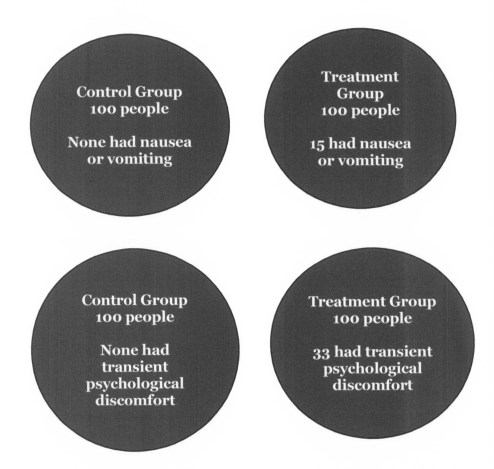

Control Group
100 people

None had nausea
or vomiting

Treatment
Group
100 people

15 had nausea
or vomiting

Control Group
100 people

None had
transient
psychological
discomfort

Treatment Group
100 people

33 had transient
psychological
discomfort

Pretty clear harms. Good job. A wise patient can discuss the potential risks of taking Magic Mushrooms, whatever they are, meaningfully with his or her physician.

Now for a poor example, **Blood Clot Removal Could Help More Stroke Victims** from ABC News.[82] Here's the closest this article comes to describing and quantifying the harms:

> In total, 1,287 patients were enrolled in the five trials studied.

> The researchers...found that the patients who received standard medical therapy along with an endovascular thrombectomy up to 7.3 hours after developing stroke symptoms were less likely than patients who were treated with only medications to report disability three months later.

That's it! No other description of potential risks from brain surgery. Brain surgery! Poorly done.

I want to include one more example, this time of disease risks because people sometimes learn of a disease risk from an article or report, think 'that might be me' and change their behavior or look to a medical treatment as a result.

We'll consider the December 14, 2018 New York Times article **Is Eating Deli Meats Really That Bad for You?** [83]

This article quoted Dr. Nigel Brockton of the American Institute for Cancer Research as saying "We see a 4 percent increase in the risk of cancer even at 15 grams a day, which is a single slice of ham on a sandwich."

- 4% of what?
- Over what time period?
- For what population?
- Does this apply only to people who eat 1 slice of ham each day? What happens if I eat a sandwich every week or two?

- This statement is more emotional than scientific: it purports to quantify cancer risks but doesn't actually provide useful or meaningful facts.

The article goes on to say "Eating a more typical serving of 50 grams of processed meat a day would increase the risk of colorectal cancer by 18 percent…"

- My personal Risk Chart in Chapter 3 of this book says that I have a 0.6 in 100 chance of developing colon cancer over the next 10 years. An 18% risk increase takes this to 0.7. And that's apparently only if I eat 50 grams every day.

For context and fun (fun?), I read another report from the American Institute for Cancer Research, the same organization quoted in this article, that suggests eating romaine lettuce may reduce the risk of stomach cancer.[84] So if I put romaine lettuce on my occasional ham sandwich…

That's why this is a bad article. It doesn't provide any meaningful risk analysis or context. It's more emotional than substantive. It doesn't tell the reader how this information applies to him or her.

Let's close this section of how news articles report benefits and harms:

- Beware of red flag emotional words like 'dramatic', 'promising' and 'game changer'.
- Try to put the benefits and harms into our comparative study graphic per 100 people. Good articles should provide the necessary information.
- Pay attention to expressions like '50% better than' or 'reduces your risk by 50%'. They may make medical interventions appear more impressive than they really are.
- Beware of risk analysis stated as percentages. An 18% risk increase may only be 0.1 person in 100 over 10 years.
- Look closely for treatment harms. If they aren't clearly stated, proceed cautiously.

Does the article discuss the evidence quality?

Some articles describe **animal studies**. I ignore them. Though these may be useful for scientific research, the differences between mice and people are far too huge for patients to draw any useful clinical conclusions. Move on.

Short term trials are a big red flag. A drug that benefits patients over a 6 month period might begin to harm them after a year or two. Remember the Celebrex discussion from Chapter 2. Celebrex showed lower rates of stomach and intestinal ulcers than two other drugs during an initial 6 month trial but that safety advantage disappeared over the next 6 months.

A good medical news article tells the length of time any treatments or medications were studied; a bad article doesn't. Pay attention to this. You may be prescribed a drug for years and learn that it's only been tested for months.

Beware of observational studies. We discussed this earlier in this chapter. Observational studies discover correlations, not causality. Try to remember this when you read an article. Good ones explain the type of evidence; poor ones may obfuscate. I'm always skeptical of observational study conclusions in case you hadn't noticed.

Here's a humorous (?) example of how bad observational studies can be. Consider these two quotes from articles in the Lancet a year apart:

- **High carbohydrate intake was associated with higher risk of total mortality... Total fat and types of fat were not associated with cardiovascular disease, myocardial infarction, or cardiovascular disease mortality.** Associations of fats and carbohydrate intake with cardiovascular disease and mortality, Dehghan et al, Lancet, **Aug 29, 2017**
- **Low carbohydrate dietary patterns favoring animal-derived protein and fat sources from sources such as lamb, beef, pork, and chicken, were associated with higher mortality** Dietary carbohydrate intake and

mortality, Seidelmann, The Lancet Public Health, **Aug 16, 2018**

Apparently both high and low carb diets are associated with higher mortality, and types of fat are similarly both associated with and not associated with higher mortality.

(See why I still like Cochrane and the US Preventive Services Task Force?)

These comments from Time's article **Alcohol Does a Body Good? Study Finds It Boosts Bone Health** summarize the problems with observational studies quite well.[85]

- Women who drink moderately have higher bone density than non-drinkers or heavy drinkers — but such observational studies connecting people's dietary or drinking patterns to health effects have not been able to determine cause and effect.

- It's possible, for example, that people who are healthier to begin with are more likely to drink moderately, rather than the other way around.

- While the study links drinking with improvements in bone health, it doesn't go so far as to show that alcohol can reduce women's risk of breaking a bone.

- Although there is substantial evidence that moderate alcohol consumption correlates with higher bone mass density in postmenopausal women, it is much less clear whether consuming alcohol lowers the fracture rate.

Surrogate or indicator benefits are another big red flag. We discussed these in Chapters 2 and 3. Indicators include things like cholesterol levels, blood pressure levels, blood sugar, bone density, etc. All *seem* to indicate something about health and patient outcomes but may not.

- Niaspin, the niacin-based product that increased LDL cholesterol, didn't reduce the heart attack rate.
- Ezetimibe, the LDL-cholesterol lowering drug, didn't prevent heart disease.
- Lowering blood pressure with beta blockers so it conforms to some guideline doesn't reduce your heart attack or cardiovascular mortality risk.

Wise patients seek studies that measure patient events, not indicators.

Beware of study summaries or headlines that don't accurately or usefully reflect the evidence. Consider these comments about colonoscopies:

> "one of the most effective cancer screening and prevention exams" WebMD [86]

> "As many as 60 percent of colon cancer deaths could be prevented if everyone 50 years old or above underwent colonoscopies" Dana Farber Hospital [87]

> "Colonoscopies are safe and have been proven to be an effective way to screen for colon cancer" Science Daily which calls itself 'your source for the latest research news.' [88]

But when we quantify the benefits and harms, we find something surprising. I'll show that with the comparative study graphic using my best estimates about colonoscopy benefits per **100, 50-year old non-smoking men over 10 years**. The control group mortality estimates come from the Risk Charts in Chapter 3; the treatment group estimates from various sources [89] including Dana Farber's statement above.

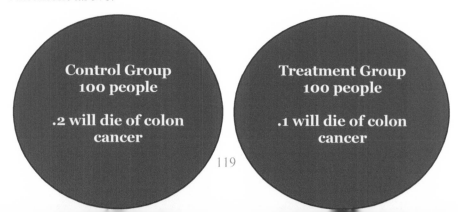

Control Group
100 people

.2 will die of colon cancer

Treatment Group
100 people

.1 will die of colon cancer

The benefit here is about 0.1 life saved per 100 men screened over 10 years.

Can you see how Dana Farber's statement that 'as many as 60% of colon cancer deaths could be prevented' might confuse readers? I used a 50% estimate above.

Again, a rule of thumb. Whenever you read that something cuts your mortality risk by x%, refer back to the Risk Charts. They're reasonable estimates.

My best guess about harms is that colonoscopies cause about .1 colon perforation per 100 patients.[90] Again, the comparative study graphic.

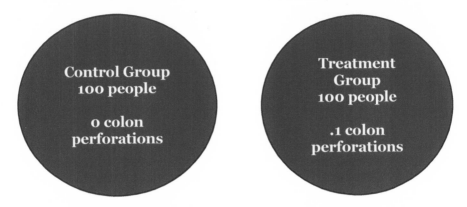

I use this example to show two things:

- First, why expressions like 'very beneficial with only infrequent harms' don't mean much. Here, the number of patients benefitting and being harmed is almost exactly the same.

- Second, why evidence quality is so important in medical news articles. Merely reiterating a 60% potential mortality reduction doesn't suffice. In this case, it's about .1 in 100 50-year-old men over 10 years. Is this clinically significant and a game changer? I don't know. You and your doctor decide.

Beware of anecdotes like 'my friend's colonoscopy saved his life' or a medical article about an individual case. Anecdote is not singular for data! Your friend's life *may* have been saved by a colonoscopy.

But he or she also could have been overdiagnosed and received unnecessary treatment.

Or the colon abnormality might have turned symptomatic in plenty of time for treatment.

Or there may be something unique about your friend that doesn't apply to most other people, including you.

We don't know from your story if the colonoscopy actually saved your friend's life or what conclusion to draw from the experience.

Some take-aways from this section:

- Don't draw clinical conclusions from animal studies.
- Try to ensure that the treatment has been studied for the same length of time that you'll use it.
- Read observational studies very skeptically. They may not show what they purport to show.
- Question surrogate or indicator benefits. How closely do they reflect actual patient medical events?
- Be cautious about relying on study or article headlines. These may or may not reflect the actual study data.

Did the authors disclose any conflicts of interest?

This at first glance appears one of those boring caveats that nerds like me worry about. But I caution you to consider why disclosing – or avoiding - author conflicts of interest is so important. It's a really big deal.

Conflicted authors may

- Ignore or discard negative information
- Inappropriately or unscientifically select the data and population studied
- Slant the writing, especially the study abstract, the only part that most people read, and
- End up with biased or corrupted conclusions.

The Journal of Legal Studies published a fascinating analysis of advisor bias that also applies to medical research in 2005.[91]

They discovered, first, that readers generally do not discount biased advice as much as they should due to the amount of bias evident in the advice. In other words, according to their analysis, people who stood to gain financially altered their advice to benefit themselves.

But second and perhaps more striking, the Journal analysis discovered that disclosing financial conflicts can actually *increase* the advice bias. Disclosure apparently allows advisors or authors to feel 'morally licensed and strategically encouraged' (their words) to exaggerate their advice even further.

Disclosure, in other words, seems to give people 'a green light to behave unethically, as it they were absolved from having to consider other's interests' according to the Boston Globe's analysis in 2011. [92]

Consider those two conclusions as context for a 2018 Journal of the American Medical Association study that found only 37% of doctors paid by medical device manufacturers disclosed those relationships in medical journal articles.[93] Readers often don't know if, or how, any given study's conclusions may be skewed by author financial concerns.

A classic example of all this is an article titled 'Why Olanzapine Beats Risperidone, Risperidone Beats Quetiapine, and Quetiapine Beats Olanzapine' published in the American Journal of Psychiatry, 2006, showing the impact of pharmaceutical funding on study conclusions.[94]

How does a wise patient proceed? Only read study articles that disclose author conflicts? Or avoid all such articles? These are difficult questions to answer, given the complexity and omnipresence of medical news articles and advisor conflicts.

My standard advice here: rely on Cochrane and the US Preventive Services Task Force.

Does the article relate new information to existing?

Medical research is ongoing and evolutionary. You, a non-medically trained patient, rarely identify groundbreaking information that you need to share with your doctor.

For every patient who says correctly 'Doc, I found this study online that completely transforms medical care', some 1000 or 10,000 or maybe 100,000 or 5 million more get a polite response like 'There are lots of studies showing basically the same thing.'

A good article helps people avoid those problems by putting the new information in context; a poor article does not.

****

Let's apply all the tools we've discussed now to a New York Times article **Fish Oil May Reduce Heart Attack and Stroke Risk for Some** by Anahad O'Connor that ran on September 25, 2018. Again, as with all the articles discussed in this chapter, I have no particular interest in the specific topic - fish oil - but a big interest in the information presentation.

First, a couple sentences as written:

> Fish oil has long been a popular supplement to protect against heart disease. It contains high levels of omega-3 fatty acids,

primarily EPA and DHA, which reduce inflammation and lower triglyceride levels...

Now my comments:

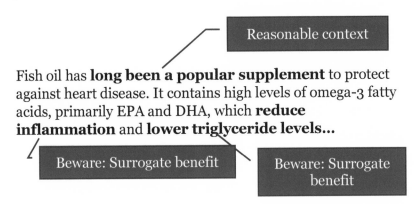

Fish oil has **long been a popular supplement** to protect against heart disease. It contains high levels of omega-3 fatty acids, primarily EPA and DHA, which **reduce inflammation** and **lower triglyceride levels...**

As written:

But until now most of the clinical trials that have looked at fish oil in heart patients had not found convincing evidence that it helps. Some argued that the trials were deeply flawed... Some of the studies were observational... They also used various types of fish oil.

My comments:

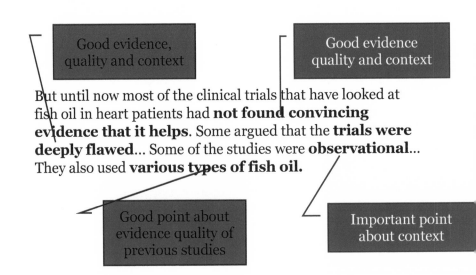

But until now most of the clinical trials that have looked at fish oil in heart patients had **not found convincing evidence that it helps**. Some argued that the **trials were deeply flawed**... Some of the studies were **observational**... They also used **various types of fish oil.**

As written:

> The intervention in this trial, which was sponsored by Amarin, was not the typical fish oil supplement that can be purchased at any supermarket or pharmacy. Vascepa is a prescription drug that contains highly purified EPA. Fish oil supplements, on the other hand, often contain a mixture of both EPA and DHA... DHA tends to raise LDL cholesterol, the so-called bad kind associated with heart disease.

My comments:

Good context

Potential conflict of interest

The intervention in this trial, which was **sponsored by Amarin**, was **not the typical fish oil supplement** that can be purchased at any supermarket or pharmacy. Vascepa is a **prescription drug that contains highly purified EPA**. Fish oil supplements, on the other hand, often contain a mixture of both EPA and DHA... **DHA tends to raise LDL cholesterol,** the so-called bad kind associated with heart disease.

Context, something potentially new here

Surrogate problems with previous supplements, context. How important is this surrogate?

As written:

> The new trial showed that statin-treated adults with elevated triglycerides who were prescribed high doses of the purified EPA had a 25 percent reduction in their relative risk of heart attacks, strokes and other cardiac events compared to a control group of patients who received placebo.

> The trial found that Vascepa was safe and well tolerated.

> Dr. Ethan Weiss, a cardiologist and associate professor at the University of California, San Francisco, who was not involved in the study, said… "really big…excited".

My comments:

How many benefits & harms / 100 patients?

The new trial showed that statin-treated adults with elevated triglycerides who were prescribed high doses of the purified EPA had a **25 percent reduction in their relative risk** of heart attacks, strokes and other cardiac events compared to a control group of patients who received placebo.

The trial found that Vascepa was **safe and well tolerated**.

Emotion

Dr. Ethan Weiss, a cardiologist and associate profess University of California, San Francisco, who was not involved in the study, said… **"really big…excited"**.

Hype, emotion

I hope you can see how attentively you have to read a medical news article to become properly informed.

Summary and conclusion
The orange juice treatment for toenail fungus [95]

Any relationship between this story and reality is entirely coincidental and unintended.

Dr. Smith's patient Marie had a history of toenail fungus. She wintered in Florida last year and drank lots of fresh orange juice.

At her next annual physical, Dr. Smith noticed that her toenail fungus had diminished. Hmmm...

Smith sent a letter about Maria to the North American Journal of Fungal Disease where Dr. Brown, a dermatologist, read it. He proceeded to ask all his patients about their orange juice consumption habits and – lo and behold – found a correlation between orange juice drinking and toenail fungus.

Brown published his findings in the Journal of Citrus Fruits which was read by Nancy, a high school student working on a homework project. Nancy's father was a vice president of the Orange, Lemon and Grapefruit Tree Growers Association of Greater Orlando.

The Orange, Lemon and Grapefruit Tree Growers Association sponsored a small observational study and learned that people who drank 2 or more glasses of orange juice daily had very little toenail fungus. This study was published in the Mid-Atlantic Journal of Medicine, a quasi-scientific newsletter that reports on natural components in medications. It received a brief mention in a national Sunday newspaper supplement under the heading 'Orange Juice May Cure Toenail Fungus' and led to a spike in orange juice sales.

That was noticed by people at the American Cooperative of Juice Producers which sponsored a 2500-person comparative study limited to people who had mild toenail fungus for less than 6 weeks. All participants continued with their current toenail fungus treatment but half also added 2 glasses of orange juice daily.

After 2 weeks, 3 people in the control group and 4 in the orange juice group reported clearer toenails.

The Amalgamated Press ran this article in their health section.

Breakthrough, All-Natural Treatment for Toenail Fungus

In a recent study, people who drank 2 glasses of orange juice daily reported 33% less toenail fungus than non-orange juice drinkers.

'This is a game changer,' said Dr. Alfred Apple, lead investigator. 'We'll never treat toenail fungus without orange juice again.'

'I never thought my toes would look this good,' said study participant Austin Powers.

Dr. James Bond of Pacific Atlantic University who was unaffiliated with the study said 'I never expected results this dramatic' adding 'I assume the orange juice was shaken, not stirred.'

Orange juice company stocks rose 8% on the news.

****

Now do you see why I generally stick to Cochrane, the US Preventive Services Task Force and ChoosingWisely?

Chapter 5: A lighthearted summary

Sometimes in jest we make insightful comments.

In this closing chapter, I'll introduce some timeless, pithy and sarcastic observations about medical professionals – I hope they make you smile.

But I have a serious intent.

That so many observers have suggested the same things, over so many years, makes me wonder how true these humorous comments really are, and how descriptive they may be.

And as you chuckle at the humor, ask yourself about the patient's role in all this. How might the quotes be different if patients had been really well informed when physicians tried to treat them? How might asking the right questions have changed these observations?

I'll first list some general observations, then focus on George Bernard Shaw's 1904 play The Doctor's Dilemma. Shaw overstates his case (I hope).

You can decide for yourself how relevant these observations are today.

Some general observations: [96]

> God heals and the doctor takes the fees. *Benjamin Franklin*
>
> More people die from the cure than the disease. *Moliere*
>
> He must have killed a good many people to have made himself so rich. *Moliere, referring to a wealthy doctor.*
>
> The best doctors in the world are Doctor Diet, Doctor Quiet and Doctor Merryman. *Jonathan Swift*
>
> One of the first duties of the physician is to educate the masses not to take medicine. *William Osler, a founder of Johns Hopkins Medical School and often called the father of modern medicine.*
>
> The art of medicine consists of amusing the patient while nature cures the disease. *Voltaire*
>
> I firmly believe that if the whole *materia medica* could be sunk to the bottom of the sea, it would be all the better for mankind and all the worse for the fishes. *Oliver Wendell Holmes*
>
> Minor surgery is when they do the operation on someone else, not you. *Bill Walton former Boston Celtic with legendary foot problems.*

Now Shaw's Doctor's Dilemma about the conflict between the doctor's profit motive and patient medical needs with some current data and observations.

That any sane nation, having observed that you could provide for the supply of bread by giving bakers a financial interest in baking for you, should go on to give a surgeon a financial interest in cutting off your leg, is enough to make one despair. But that is precisely what we have done.

- Most doctors in the US are paid fees for the services they provide. More or less like Shaw's bakers.

There is a fashion in operations as there is in sleeves and skirts. The triumph of some surgeon who has at last found out how to make a once desperate operation fairly safe is usually followed by a rage for that operation not only among the doctors, but actually among their patients.

- The US annual number of coronary angioplasty procedures rose from 3500 in 1982 to around 500,000 in 2015 at an average cost of $27,000 despite extensive research suggesting ineffectiveness:
  o A 2007 study in the New England Journal of Medicine concluded the procedure made no difference in heart attack and death rates,
  o A 2014 meta study in the Journal of the American Medical Association Internal Medicine found it provided no benefit over medical management in preventing heart attacks or death in patients with stable coronary disease,
  o A 2017 study in the Lancet concluded it doesn't reduce chest pain in stable patients.
- But we spend over $10 billion annually on it!

By making doctors tradesmen, we compel them to learn the tricks of trade.

- In 1993, there were 6 combined MD/MBA programs offered by US medical schools.

- Today there are 57. That's a 460% increase in case you like relative statistics.

Suppose, for example, a royal personage gets something wrong with his throat or has a pain in his inside. If a doctor effects some trumpery cure with a wet compress or a peppermint lozenge nobody takes the least notice of him.

But if he operates on the throat and kills the patient, or extirpates an internal organ and keeps the whole nation palpitating for days whilst the patient hovers in pain and fever between life and death, his fortune is made.

The wonder is that there is a king or queen left alive in Europe.

- There were, according to Wikipedia, 16 kings or queens in Europe when Shaw wrote his play. There are only 9 today. Makes you wonder....

I hope this book helps you avoid the problems Shaw and the others articulated so you only get the best possible care.

Notes:

---

[1] Richard Harris, Rigo Mortis and John Wennberg, Tracking Medicine for example.
[2] See the Dartmouth Atlas of Healthcare for example on this.
[3] State of Washington 2018 report First Do No Harm. I used this source for the other examples in this section also.
[4] Wennberg, Tracking Medicine. He estimates that patients have options about 85% of the time.
[5] See the Dartmouth Atlas of Healthcare and various research papers from the Dartmouth Institute for Health Policy and Clinical Practice, for example. Also David Cutler's estimate in The Quality Cure, page 20.
[6] See Wennberg, Tracking Medicine, Chapter 1

[7] HHS, Quick Guide to Health Literacy,
https://health.gov/communication/literacy/quickguide/factsbasic.htm

[8] Mulley, et al, Patient Preferences Matter, Kings Fund and the Dartmouth Center for Health Care Delivery Science, 2012, page 9

[9] This specific quote comes from Sheri Fink's summary of Atul Gawande's thoughts in her review of Gawande's book Being Mortal, NY Times Book Review, Nov 6, 2014

[10] Information on their Form 10-K, filed 3/1/17 for the period ending 12/31/16, pages 2 and 61.
http://www.annualreports.com/HostedData/AnnualReports/PDF/NASDAQ_WBMD_2016.pdf

[11] https://www.cochrane.org/news/cochrane-announces-support-new-donor

[12] https://guides.mclibrary.duke.edu/ebmtutorial/selecting-a-resource

[13] Harvey, Summaries of Cochran Systematic Reviews, Nature, February 23, 2018

[14] https://www.amazon.com/The-Cochrane-Collaboration-Medicines-Best-Kept/dp/1927755301

[15] Quoted on Full Measure with Sheryl Attkisson, August 6, 2017

[16] https://www.statista.com/statistics/781756/atenolol-prescriptions-number-in-the-us/

[17] https://www.ashp.org/drug-shortages/current-shortages/Drug-Shortage-Detail.aspx?id=334

[18] Estimate comes from David Bach's summary, Cardiac Stress Test Trends Among US Patients < 65 Years, in American College of Cardiology, Nov 15, 2016
https://www.acc.org/latest-in-cardiology/journal-scans/2016/11/14/11/45/tuesday-140pm-et-cardiac-stress-test-trends-aha-2016
and from my own back-of-the-envelop estimate based on the State of Washington's First Do No Harm study discussed later in this chapter.

[19] See, for example, Guirguis-Blake et al, Current processes of the US Preventive Services Task Force, Annals of Internal Medicine, July 17, 2007 and 'US Preventive Services Task Force: The gold standard of evidence-based prevention' by Dr. Doug Campos-Outcalt in The Journal of Family Practice https://www.mdedge.com/jfponline/article/60333/us-preventive-services-task-force-gold-standard-evidence-based-prevention#. Among the groups explicitly stating 'gold standard' are the American Academy of Family Physicians, The Society of Teachers of Family Medicine, The Association of Departments of Family Medicine, The North American Primary Care Research Group and America's Health Insurance Plans, the largest trade association of health insurance companies with over 1300 members

[20] These are my easy-to-understand grade summaries, not quotes from the USPSTF.

[21]
https://www.uspreventiveservicestaskforce.org/Page/Document/RecommendationStatementFinal/vitamin-d-calcium-or-combined-supplementation-for-the-primary-prevention-of-fractures-in-adults-preventive-medication

22 Hojat et al, Physician's empathy and clinical outcomes for diabetic patients, Academic Medicine, March 2011, https://www.ncbi.nlm.nih.gov/pubmed/21248604 , Decety et al, Why empathy has a beneficial impact, Frontiers in Behavioral Neuroscience, January 2015 https://www.ncbi.nlm.nih.gov/pmc/articles/PMC4294163/

23 Pollak et al, Physician empathy and listening, Journal of the American Board of Family Medicine, November 2011, https://www.ncbi.nlm.nih.gov/pmc/articles/PMC3363295/

24 Quick Guide to Health Literacy, US Department of Health and Human Services, https://health.gov/communication/literacy/quickguide/factsbasic.htm

25 Brawley, How We Do Harm, page 243

26 FDA approves new label changes and dosing for zolpidem, https://www.fda.gov/Drugs/DrugSafety/ucm352085.htm

27 Armstrong, Abbott Doubled Niaspin US Sales Before Trials Cut Use, Bloomberg, June 10, 2013

28 This sentence paraphrases the New England Journal of Medicine discussion of the AIM High study http://www.nejm.org/doi/full/10.1056/NEJMoa1107579#t=article.

29 See Merck's Press Release of March 9, 2013 which includes the sentence 'Adding TREDAPTIVE to statin therapy did not significantly further reduce the risk of major vascular events compared to statin therapy in patients at high risk of cardiovascular events.' See CBS News 'Heart Drug Tredaptive is Ineffective', Jonathan Lapook, July 29, 2013 for a short summary.

30 See, for example, the ENHANCE study, Zetia's ads in Parade Magazine on Sept 11, 2011 and Zetia.com from about 2010 to 2016 when it was changed.

31 Parade Magazine, September 11, 2011. I have copies of the Sunday Seattle Times and Boston Sunday Globe. The same phrase appeared on Zetia.com for several years until the site changed in 2016 or 2017.

32 For a good summary of those studies, with expanded comments, see Sham-Wow by Walter Eisner in Orthopedics This Week, August 11, 2009, https://ryortho.com/2009/08/sham-wow/

33 North American vertebroplasty market to be slashed by over 50%, GobalData Healthcare, May 3, 2017 https://www.globaldata.com/north-american-vertebroplasty-market-to-be-slashed-by-over-50/

34 Sherman, Top doc urges restraint, Reuters, October 4, 2007 https://www.reuters.com/article/us-doctor-technology/top-doc-urges-restraint-in-adopting-new-technology-idUSN0436767620071004

35 Estimate from The Burden of Musculoskeletal Diseases in the United States, 2014 edition https://www.boneandjointburden.org/2014-report/iie0/spine-procedures

36 https://www.toyourhealth.com/mpacms/tyh/article.php?id=1447

37 Kumar and Nash, Healthcare Myth Busters, Scientific American, March 25, 2011

38 Harris, Before Stating a Statin, Talk It Over with Your Doctor, NPR All Things Considered, December 3, 2018

39 Roberts, The Trust About Statins, page 50. That estimate was 10%. I reduced it

here to be conservative.

[40] Roberts, The Truth About Statins, page 56

[41] Sattar, Statins and the Risk of Incident Diabetes, Lancet, Feb 27, 2010

[42] Okie, Missing Data on Celebrex, Washington Post, Aug 5, 2001 and Drug Firm May Not Call Celebex Safer, Washginton Post June 8, 2002

[43] Angell, The Truth about Drug Companies, page 108 (Celebrex) and 112 (antidepressants)

[44] See the American Cancer Society's Cancer Facts and Figures books to compare breast cancer incidence and mortality rates in both states by year.

[45] Clemens and Gottlieb, Do physician's financial incentives affect medical treatment https://papers.ssrn.com/sol3/papers.cfm?abstract_id=2101251

[46] Rosenthal, An American Sickness, Chapter 3, page 113 in the large print edition

[47] Rosenthal, An American Sickness, Chapter 3, page 106 in the large print edition.

[48] Mulley, et. al. Patient Preferences Matter: Stop the Silent Misdiagnosis, Kings Fund and the Dartmouth Center for Health Care Delivery Science, 2012

[49] Landro, Weighty choices in patient's hands, Wall Street Journal, August 4, 2009

[50] Porter and Teisberg, Redefining Health Care page 60

[51] This summary from Crist 'Don't pick doctors based on where they went to medical school' Reuters, October 9, 2018 based a study published in the BMJ online Sept 26, 2018

[52] Bakalar, Choose a thyroid surgeon who does dozens of operations per year, NY Times, March 16, 2016

[53] Gawande, ComplicationsL a surgeon's notes on an imperfect science. These comments come from the chapter titled 'The computer and the hernia factory' in Part 1.

[54] Urbach, Pledging to Eliminate Low Volume Surgery, NEJM, Oct 8, 2015

[55] Gawande, Complications, chapter entitled Education of a Knife.

[56] The term 'overdiagnosis' was perhaps coined by Dr. H. Gilbert Welch in his excellent book Overdiganosed: Making People Sick in Pursuit of Health. The discussion above comes largely from Kale et al, Overdiagnosis in primary care, BMJ, August 14, 2018

[57] Be sure to understand the Disclaimers: The charts do not account for some individual characteristics that affect the chance of death, most importantly smoking. Smoking substantially increases the chance of dying from heart attacks, stroke, lung cancer, chronic lung disease, and all-causes combined. If you smoke, your chances of dying from these causes are higher than those shown in the charts – and if you never smoked, your chances of dying are lower.

Rules used by the National Center for Disease Statistics (NCHS) may result in over- and under-counting of some underlying causes of death. Diabetes and high blood pressure deaths, for example, are probably under-counted: they are often reported as contributing factors rather than as the underlying cause of death because of uncertainty in the chain of events leading to death. Flu deaths are probably over-counted since many pneumonia deaths, completely unrelated to the flu, are nevertheless attributed to it. Because the conditions are difficult to disentangle, we present a combined flu/pneumonia category.

[58] USPSTF Final Recommendation Statement, High Blood Pressure in Adults: Screening. https://www.uspreventiveservicestaskforce.org/Page/Document/RecommendationStatementFinal/high-blood-pressure-in-adults-screening

[59] https://www.ahajournals.org/doi/pdf/10.1161/CIR.0000000000000625, For an interesting discussion, see Klodas, One Cardiologist's Mission to Reduce Statin Use for Cholesterol, CNN.com, January 9, 2019 https://www.cnn.com/2019/01/08/health/cardiologist-statin-cholesterol-mission/index.html

[60] US Department of Health and Human Services, National Institutes of Health, National Heart, Lung and Blood Institute https://www.nhlbi.nih.gov/files/docs/public/heart/chol_tlc.pdf

[61] Guideline for Treating Blood Cholesterol to Reduce Cardiovascular Risk, https://www.cardiosmart.org/heart-conditions/guidelines/cholesterol

[62] See the note on the bottom right http://chd.bestsciencemedicine.com/calc2.html

[63] Han, Effect of statin treatment, JAMA Internal Medicine, May 31, 2017

[64] These come from Jain, Can a physician write a prescription for friendship?, Harvard Medical Magazine, Autumn, 2018

[65] Summary from Jauhar, Heart: A History, page 24

[66] Work and the Loneliness Epidemic, Harvard Business Review, September 27, 2017

[67] The British Whitehall studies explored this issue in depth.

[68] Class, the ignored determinant of health, New England Journal of Medicine, September 9, 2004

[69] The British Whitehall studies examine these issues in depth.

[70] Drexler, The People's Epidemiologist, Harvard Magazine, March 2006

[71] Steven Hatch, Snowball in a Blizzard, 2016

[72] I got the idea for this example from Welch in Overdiagnosed.

[73] Kolata, Sports Medicine Said to Overuse MRI's, New York Times, October 28, 2011

[74] Wruble TED talk. Here's the link courtesy of the Lown Institute https://lowninstitute.org/news/looking-past-statistical-magic-tricks-dr-jill-wruble-cancer-screening/

[75] This slide came from Dr. Wruble's TED talk. https://lowninstitute.org/news/looking-past-statistical-magic-tricks-dr-jill-wruble-cancer-screening/

[76] Welch, Income and cancer overdiagnosis, NEJM, June 8, 2017

[77] https://www.cnn.com/2018/10/19/health/study-not-exercising-worse-than-smoking/index.html

[78] Grady, Taking Fish Oil During Pregnancy is Found to Lower Child's Asthma Risk, New York times, Dec 28, 2016

[79] Begley, In a Small Alzheimer's study, hints of modest benefit, STAT, May 1, 2017

[80] Healy, PSA Screening for Prostate Cancer Saves Lives After All, Study Says. Los Angeles Times, September 4, 2017

[81] Healy, There's something in magic mushrooms that's shown to ease anxiety and depression in cancer patients in one dose, Los Angeles Times, November 30, 2016

[82] Mohney, Blood Clot Removal Could Help More Stroke Victims, Study Finds, ABCNews.go.com, Sept 27,2016

[83] https://www.nytimes.com/2018/12/14/well/eat/is-eating-deli-meats-really-that-bad-for-you.html

[84] AICR's Foods That Fight Cancer™, Dark Green Leafy Vegetables http://www.aicr.org/foods-that-fight-cancer/foodsthatfightcancer_leafy_vegetables.html

[85] Szalavitz, Alcohol Does a Body Good? Study Finds It Boosts Bone Health, Time, July 11, 2012

[86] https://www.webmd.com/colorectal-cancer/news/20110103/new-evidence-on-benefits-of-colonoscopies#1

[87] https://blog.dana-farber.org/insight/2013/03/should-i-get-a-colonoscopy/

[88] https://www.sciencedaily.com/releases/2017/03/170315144501.htm

[89] Colonoscopy harms: Complications of Colonoscopy, American Society for Gastrointestinal Endoscopy 2011 https://www.asge.org/docs/default-source/education/practice_guidelines/doc-56321364-c4d8-4742-8158-55b6bef2a568.pdf?sfvrsn=8

- Johns Hopkins Medicine Colorectal Cancer website 1/24/14 http://www.hopkinscoloncancercenter.org/CMS/CMS_Page.aspx?Current UDV=59&CMS_Page_ID=33CD25B0-CCC6-4F55-A226-3C202E67D0B1

- Treatment group mortality rate: Zauber et al, Colonoscopic Polypectomy and Long-Term Prevention of Colorectal-Cancer Deaths, New England Journal of Medicine, February 23, 2012, easy to read summary in the New York Times, *Report Affirms Live Saving Role of Colonoscopy*, Denise Grady, February 22, 2012

[90] Ibid.

[91] Cain et al, The Dirt on Coming Clean: Perverse Effects of Disclosing Conflicts of Interest, Journal of Legal Studies, Jan 2005

[92] Humphries, Deeply Conflicted, Boston Globe May 15, 2011

[93] Kowalczyk, Doctors warned on industry ties, Boston Globe, Oct 10, 2018

[94] Heres et al, Why Olanzapine Beats Risperidone, Risperidone Beats Quetiapine, and Quetiapine Beats Olanzapine, American Journal of Psychiatry, Feb 1, 2006

[95] Thanks to Gary Schwitzer in his book Covering Medical Research for most of the ideas in this case study.

[96] I got these quotes from various sources including notable-quotes.com

http://www.notable-quotes.com/d/doctors_quotes.html, quotegarden.com
http://www.quotegarden.com/medical.html,